Soil Investigation and Foundations Design

Soil Investigation and Foundations Design

Prepared By:

Mohamed A. El-Reedy, Ph.D.

2020 by Mohamed A. El-Reedy

No claim to original U.S. Government works

This book contains information obtained from authentic and highly regarded sources. Reasonable efforts have been made to publish reliable data and information, but the author and publisher cannot assume responsibility for the validity of all materials or the consequences of their use. The authors and publishers have attempted to trace the copyright holders of all material reproduced in this publication and apologize to copyright holders if permission to publish in this form has not been obtained. If any copyright material has not been acknowledged please write and let us know so we may rectify in any future reprint.

Except as permitted under U.S. Copyright Law, no part of this book may be reprinted, reproduced, transmitted, or utilized in any form by any electronic, mechanical, or other means, now known or hereafter invented, including photocopying, microfilming, and recording, or in any information storage or retrieval system, without written permission from the author.

For permission to photocopy or use material electronically from this work, please access or contact the Copyright Clearance Center, Inc. (CCC), 222 Rosewood Drive, Danvers, MA 01923, 978-750-8400. CCC is a not-for-profit organization that provides licenses and registration for a variety of users. For organizations that have been granted a photocopy license by the CCC, a separate system of payment has been arranged.

Welcome to visit our website www.elreedyman.com

This book dedicated to spirit of my mother and my father; and dedicated to my wife and my son, Hisham and my daughters, Maey and Mayar

Contents

Preface 9
1 Soil Investigation 11
   1.1 Introduction 11
   1.2 Soil classifications 15
   1.3 Soil investigation specifications 22
2. Soil Exploration methods 26
   2.1 Introduction 26
   .2.2 Planning the program 27
   .2.3 Organization of field work 29
   2.4 Soil boring methods 31
   2.4.1 Wash borings 32
   2.4.2 Sampling methods: 32
   .2.5 Standard Penetration Test (SPT) 37
   2.6 Cone Penetration Tests (CPT) 41
   2.7 VANE TEST 47
   .2.8 Cross hole Test 51
   2.9 Plate Load Test 55
3 Soil investigation report 61
   3.1 Soil investigation SOW 61
   3.2 Soil investigation report 69
4. Foundation Design 71
   4.1 Shallow Foundation 71
   4.2 Deep Foundation 81
      4.2.2 Steel piles 85

- 4.2.3 Concrete piles 86
- 4.2.4 Pre-stressed Pile 88
- 4.2.6 Pile caps 97
- 5 Retaining wall 101
  - 5.1 Introduction 101
  - 5.2 Preliminary retaining wall dimensions 102
  - 5.3 Check stability against overturning 104
  - 5.4 Check stability against sliding 106
  - 5.4 Check stability against bearing capacity 108
- 6. Soil With Problems 121
  - .6.1 Dynamic compaction 121
  - 6.2 Vibro Compaction 124
  - 6.3 Compaction Grouting 126
  - 6.4 Wick Drain 128
- References: 131

The Author

Mohamed A. El-Reedy's background is in structural engineering. His main area of research is the reliability of concrete and steel structures. He has provided consulting to different engineering companies and oil and gas industries in Egypt and to international companies such as the International Egyptian Oil Company (IEOC) and British Petroleum (BP). Moreover, he provides different concrete and steel structure design packages for residential buildings, warehouses, and telecommunication towers and electrical projects with WorleyParsons Egypt. He has participated in Liquified Natural Gas (LNG) and Natural Gas Liquid (NGL) projects with international engineering firms. Currently, Dr. El-Reedy is responsible for reliability, inspection, and maintenance strategy for onshore concrete structures and offshore steel structure platforms. He has performed these tasks for hundreds of structures in the Gulf of Suez in the Red Sea.

Dr. El-Reedy has consulted with and trained executives at many organizations, including the Arabian American Oil Company (ARAMCO), bp, Apachi, Abu Dhabi Marine Operating Company (ADMA), the Abu Dhabi National Oil Company, King Saudi's Interior ministry, Qatar Telecom, the Egyptian General Petroleum Corporation, Saudi Arabia Basic Industries Corporation (SAPIC), the Kuwait Petroleum Corporation, and Qatar Petrochemical Company (QAPCO). He has taught technical courses about repair and maintenance for reinforced concrete structures and the advanced materials in concrete industry worldwide, especially in the Middle East.

Dr. El-Reedy has written numerous publications and has presented many papers at local and international conferences sponsored by the American Society of Civil Engineers, the American Society of Mechanical Engineers, the American Concrete Institute, the American Society for Testing and Materials, and the American Petroleum Institute. He has published many research papers in international technical journals and has authored ten books about quality management and quality assurance, economic management for engineering projects, and repair and protection of reinforced concrete structures, offshore structure and reliability for RC structures. He received his bachelor's degree from Cairo University in 1990, his master's degree in 1995, and his Ph.D from Cairo University in 2000.

Preface

Any civil or structure engineer performing a construction project should deal with the soil investigation and select the best foundation of the project. So from preliminary engineering, detail and construction phase should have an interface with a geotechnical study start from preparing the scope of work for this study, receive the report and review it and perform its recommendation in our engineering and follow up during construction. The soil mechanical and its properties have its own parameters and definition that we need to understand it precisely.

This book provides the main principal and overview about soil investigation, geotechnical studies and the design of different foundations types with example and free software tools.

Therefore, it is considered as a handbook for soil investigation and foundation design, the scope of work preparation by define the required tests and number of bore holes and its depth for any project is presented herewith checklist. On the other side, there are a guideline for the main element of the report that expected to see in the soil investigation reports.

I try for this book to be simple and consider the applicable guideline for soil investigation and foundation design. So, it will be presented from practical point of view with strong technical background. This book will present the main element for design the shallow and deep foundation and there is a complete chapter about the retaining wall design.

This book illustrates everything that we need on the site when dealing with the geotechnical study without going in detail for the soil properties technical issue at the end, we need to design a safe structure durable along its life time.

In some time, we are facing a bad soil in our project, in this case it is required to enhance it to be match with our project requirement, therefore all the method that are used in case with soil with problem will be discussed in the chapter six.

This book is a hand book for civil. structural engineering and also for any engineer or project manager who is dealing with the soil investigation and foundation which is mandatory to any project. The book is associated with excel sheet which can be used in foundation design.

If you have any inquiry or need specific topics, do not hesitate to contact me and ask about the webinars for this subject.

Dr. Mohamed A. El-Reedy

Email:elreedyma@elreedyman.com

http://www.elreedyman.com

1 Soil Investigation

1.1 Introduction

Site investigation is considered the first essential operation to be performed in the analysis, design, and choice of a foundation. Such investigation provide the necessary information about geological, physical, and geotechnical properties of soil which is prerequisites to a save and economical construction of all civil engineering works.

Insufficient or in adequate information which respect to the character and bearing capacity of the underlying soil beneath and engineering structure may result in serious damage or collapse of the structure.

All structures are relying on our ability to design foundations. One of the worst failures is the failure of the Transcona Grain Elevator in 1913 Within 24 hours after loading the grain elevator at a rate of about 1 m of grain height per day, the bin house began to tilt and settle.

No borings were done to identify the soils, only , an open pit about 4 m deep was made for the foundations and a plate was loaded to determine the bearing strength of the soil.

The information gathered from the Transcona failure and the subsequent detailed soil investigation was used (Peck and Bryant, 1953; Skempton, 1951) to verify the theoretical soil bearing strength.

Peck and Bryant found that the applied pressure from loads imposed by the bin house and the grains was nearly equal to maximum pressure that the soil could withstand, thereby lending support to the theory for calculating the bearing strength of soft clay soils.

The Transcona was designed at a time when soil mechanics was not yet born. One eyewitness (White, 1953) wrote: "Soil Mechanics as a special science had hardly begun at that time. We have come a long way in understanding soil behavior since the founding of soil mechanics by Terzaghi in 1925.

Soil is a natural body comprised of solids (minerals and organic matter), liquid, and gases that occurs on the land surface, occupies space, and is characterized by one or both of the following: horizons, or layers, that are distinguishable from the initial material as a result of additions, losses, transfers, and transformations of energy and matter or the ability to support rooted plants in a natural environment.

Figure 1 Settlement in fence wall

Figure 2 Settlement in concrete Skelton

Soil is formed over a long period of time. The formation of soil happens over a very long period of time. It can take 1000 years or more. Soil is formed from the weathering of rocks and minerals. The surface rocks break down into smaller pieces through a process of weathering and is then mixed with moss and organic matter.

Geotechnical engineering is the branch of civil engineering. concerned with the engineering behavior of soil. Geotechnical engineering includes investigating existing subsurface conditions; assessing risks posed by site conditions; designing earthworks and structure foundations; and monitoring site conditions, earthwork and foundation construction.

A typical geotechnical engineering project begins with a site investigation of soil, rock, fault distribution and bedrock properties on and below an area of interest to determine their engineering properties.

Site investigations are needed to gain an understanding of the area in or on which the engineering will take place. Investigations can include the assessment of the risk to humans, property and the environment from natural hazards such as earthquakes, landslides, sinkholes, soil liquefaction, debris flows and rock falls.

In general, the purpose of making site investigation can be summarized as:

- a- To provide the civil engineer with reliable and detailed information about soil and ground water condition for the selection of the most efficient and economical type of foundation.
- b- To plan for the suitable method of construction and the efficient equipment to be used

c- To investigate whether any aggressive subsoil water or organic impurities exists in the soil.

d- To study any further condition that may affect the structure (ground water level, future excavation, etc…)

Because soil are very important construction material in which ,or on which, or by means of which civil engineering's build structures, it is doubtful if any major civil engineering works are constructed at present without site exploration being under taken. Soil investigation should provide data on the following items:

a- location of the ground water level

b- bearing capacity of soil

c- settlement predictions

d- selection of alternative depths of foundation

e- data of earth pressures on retaining structures and excavation supports.

f- data necessary for construction purposes (permeability, stratification, etc…)

1.2 Soil classifications

To classified the soil into a group according to the soil behavior and physical shape

- TYPE OF CLASSIFICATION:
    - CLASSIFICATION BY VISUAL
    - AASHTO (American Association of State Highway and Transportation Officials)
    - UCS (Unified Soil Classification)
- SOIL TESTS
    - ATTERBERG LIMIT
    - SIEVE ANALYSIS
    - HYDROMETER ANALYSIS

Atterberg Limit

- Cohesive Soil
- Base on water content
- Consistency Limit : Liquid Limit, Plastic Limit and Shrinkage Limit

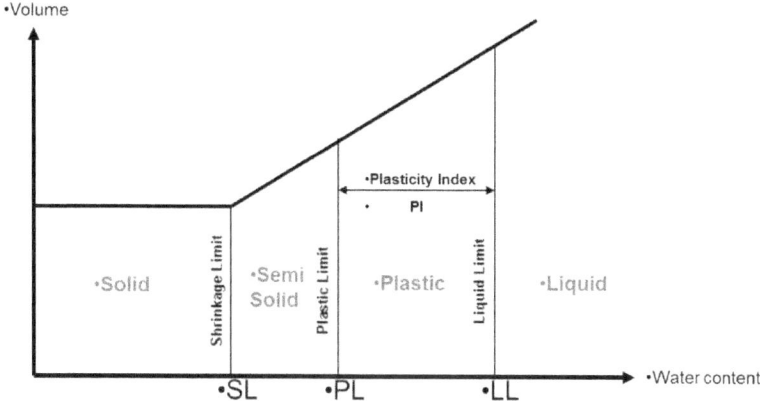

Figure 3 Relation between water content and LL, PL, SL

Liquid limit (LL)

The liquid limit is that moisture content at which a soil changes from the liquid state to the plastic state. It along with the plastic limit provides a means of soil classification as well as being useful in determining other soil properties

Two main methods to determine the liquid limit :

1. Cone Pentrometer Method
2. Casagrande Method

There are a videos for all the tests, you can contact us through www.elreedyman.com to send you the videos

Figure 4 Casagrande soil test

Plastic Limit PL

This test presents the Plastic behavior of the soil. The test is done by rolling up the soil sample to 3.2mm diameter by hand and defined as the water content, in percent, at which the soil crumbles, when rolled into threads of 1/8 in (3.2mm) in diameter.

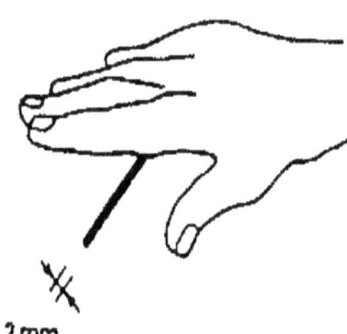

Figure 5 Rolling the soil by hand

CONSISTENCY RELATIONSHIP

- Plasticity Index (PI)

$$PI = LL - PL \quad (1)$$

- Liquidity Index (LI)

$$LI = (w-PL)/(LL-PL) \quad (2)$$

- Consistency Index (CI)

$$CI = (LL-w)/(LL-PL) \quad (3)$$

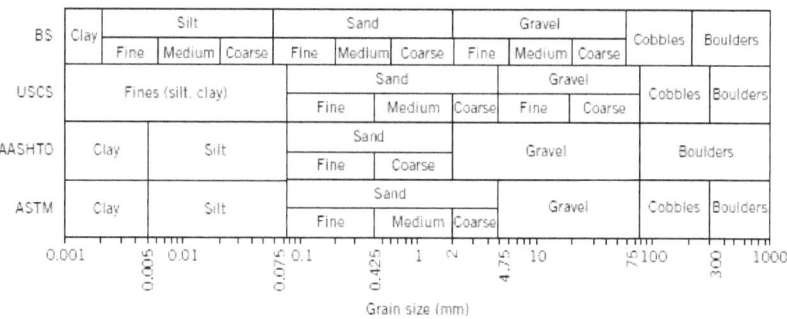

Figure 6 Soil classification based on different standard

Table 1. Soil types and sizes based on ASTM D 2487		
Soil type	Description	Average grain size
Gravel	Rounded and /or angular bulky hard rock, coarsely	Coarse: 75 to 19

	divided	mm
		Fine: 19mm to 4.75 mm
Sand	Rounded or angular hard rock	Coarse: 4.75 to 2.0 mm
		Medium: 2.0 to 0.425
		Fine: 0.425 mm to 0.075 mm
Silt	Particle size between clay and sand but it has little or no strength when dry	0.075 to 0.002 mm
Clay	Particles are smooth, has a significant strength when dry but water reduces its strength	<0.002 mm

Figure 7 Sieve analysis

Table 2   The Sieve analysis Pass percentage calculation

Size of sieve (mm)	Weight remain	Total remaining weight	Percentage of remaining	Percentage of pass
37.5	W_1	W_1	$R_1=W_1/W \times 100$	$P_1=100-R_1$
20	W_2	W_1+W_2	$R_2=W_1+W_2/W \times 100$	$P_2=100-R_2$
10	W_3	$W_1+W_2+W_3$	$R_3=W_1+W_2+W_3/W \times 100$	$P_3=100-R_3$

1.3 Soil investigation specifications

The following is a guideline in defining the scope of the soil investigation, MIN. DEPTH OF SOIL INVESTIGATION with the spacing will be as follow:

- Shallow Foundation : 3 x Foundation width (min. 9m)
- Raft Foundation   : 2 x Foundation width
- Pile Foundation   : 2 x Pile width (measured from pile tip)
- Pile + Raft Foundation : 2 x building width
- Retaining Earth Structure : 0.7 x cutting width or 1 x cutting height (take the biggest)
- Soil Embankment : 2 x embankment width
- Take on the location of the foundation
- For area 300 m2 not less than one boring
- Boring depth 10-15 m depend on soil

- Spacing between borings in multistory building, 15-30 m
- One story manufacturing plants, 30-90 m
- Define if there is silty soil or not

  NO. OF SOIL INVESTIGATION in case of initial Initial Investigation :
- Normal Soil : every 100 to 200 m
- Soft Soil : every 50 to 100 m
- Detail Investigation :
- Square structure : every 15 to 25 m
- Strip structure : every 25 to 50 m
- At the important side of the structure, the number of soil investigation can be increased

The number of boreholes should be adequate to detect variations of the soils at the site. If the locations of the loads on the footprint of the structure are known (this is often not the case), you should consider drilling at least one borehole at the location of the heaviest load.

A minimum of three boreholes should be drilled for a building area of about 250 m² (2500 ft²) and about five for a building area of about 1000 m² (10,000 ft²).

Table 3. Guideline Min. Number of boreholes in building	
Area (m2)	No. of bore holes
Less 100	2

250	3
500	4
1000	5
2000	6
5000	7
6000	8
8000	9
10,000	10

In compressible soils such as clays, the borings should penetrate to at least between 1 and 3 times the width of the proposed foundation below the depth of embedment or until the stress increment due to the heaviest foundation load is less than 10%, whichever is greater.

- In very stiff clays and dense, coarse-grained soils, borings should penetrate 5 m to 6 m to prove that the thickness of the stratum is adequate.
- Borings should penetrate at least 3 m into rock.
- Borings must penetrate below any fills or very soft deposits below the proposed structure.

- The minimum depth of boreholes should be 6 m unless bedrock or very dense material is encountered

Table 4. Guideline for bore holes number and depth

Foundation type	Min. no. of boreholes	Min. depth
Shallow foundation	1 but placed at node gride size 15 x15 m or 40x40 m	5m 0r 1 to 3 times the foundation width
Pile foundation	Sam as shallow	25 to 30 m; for bedrock drill 3m into it
Retaining wall	1 for every length 30m or 1 to 2 times the retaining wall height	1 to 2 times the wall heights, for bedrock 3 m inside it
Cut slope	1 every 60m along slope length	6m below the bottom of the cut slope
Embankment for high way	1 every 60 m	The greater of 6m or 2 times the height

2. Soil Exploration methods

2.1 Introduction

Various methods of subsurface investigation are known and widely used to be obtained a considerable amount of information about the subsoil. Among these methods: test pits and open cuts, borings (wash boring, auger boring, rotary and percussion drilling), soundings, load tests, and geophysical methods.

Some method will obviously yield more information than others and soil engineer should constantly keep in mind when planning the exploration program, the purpose of the project and the relative costs involved. Two main points, in particular, are to be considered in soil investigation

a- The size of the project if large project geological studies will be necessary to determine the type and the extend of other methods of investigations, if a small projects, some borings and the study of the neighboring foundation may be sufficient.

b- The type of structure and its sensitivity to future settlement. This study in many cases required the complete change of the statically system of the superstructure.

The choice of the exploration method is also determined from the location of the site. For compacted site in town where local information are available, borings and soundings will provide the required data. For new sites, site survey as well as geological studies are necessary to determine the program for soil investigation.

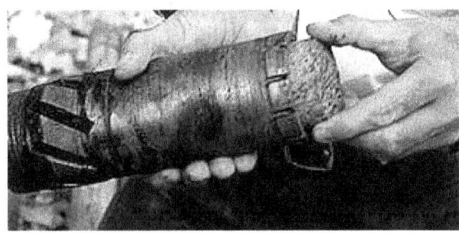

Figure 8 Soil core from boring

Figure 9 soil boring on site

.2.2 Planning the program

The actual planning of a subsurface exploration program includes some or all of the following steps.

a- assembly of all a valuable information (dimensions, columns spacing, type and use of structure, basement requirement, bridge span and pier loadings, heights or retaining structure, etc....)

b- reconnaissance of the area : field trip to the site to examine topography, type and behavior of adjacent structures such as cracks or noticeable settlement, erosion in existing cuts and soil stratification the reconnaissance may also be in the form of a study of geological maps or aerial photographs.

c- preliminary site investigation; few borings or a test pit to establish types of soil, stratification and possibly location of G.W.T.

d- detailed site investigation: for complex project and/or erratic soil conditions, samples are collected for shear strength determination and settlement analysis.

Figure 10 Open bit soil boring

.2.3 Organization of field work

After the study of the geological data, maps, and photographs, the engineer, can considered the number, type, and depth of exploratory borings required. The engineers in charge of soil investigation have to require about the results as soon as any samples are extracted and does not wait until all borings have been completed, otherwise considerable waste of money and efforts (for the chances are high that repetitive information have been collected while some data are still lacking).

To obtain the most useful information of minimum cost and effort, the soil engineer in charge of an investigation should follow a number of sample rulers as follows:

1-you must know exactly what kind of information the investigation is expected to provide. This means that he must be completely familiar with the type of construction planed on the site if the investigation concerns a proposed building, he must know exactly the value and sort of loads on the foundation the evaluation of the lowest floor level in relation to the existing ground surface. If future extensions or additional stories are planned he should know their location and loads. He should know if there are social requirements regarding allowable total or differential settlements.

2- he must write down minimum requirements for bore holes spacing and depth while taking into consideration the type of structure and estimated soil condition. The borings should provide sufficient data to enable reliable soil profile along two principals, access to be drawn. The minimum depth of boring should be related to the extent to which the subsoil will be affected by the proposed construction. This can be evaluated by using the concepts of stress distribution in soils. A useful approximate rule to determine the depth of investigation is to find the depth at which the increased in vertical stress caused by the purposed structure will be less the 10 % of the originally existing vertical stress.

For cuttings, retaining walls and other structures where the lateral stability is the main concern, the investigation should extend below the depth of the deepest possible surface of failure. In soils the frequency or the sampling is usual governed by maximum allowable spacing of 1-1.5 m plus the condition that sample must be taking each change in soil conditions, investigation for dams, reservoir and landslides demand continuous sampling is obtained by a special apparatus.

3- The engineer must select the proper drilling and samp0ling equipment with attention to local preference for certain types of equipment which exists through out the worlds

4- the engineer should select the field supervisor and must be apple to layout boring locations with reasonable accuracy.

2.4 Soil boring methods

The main purpose of drilling a test hole or boring is to obtain samples of the various strata and to acquire an accurate picture of the subsoil profile at one location. Different methods of soil boring are available and will be discussed in the following:

1- Test pits and open cuts.

This method is ideal for soils which required little support for excavation. This test hole is made large enough to permit an observer to descend to log the soil profile visually and the succession of strata can be examined in the wall of the4 pit. Block samples may be taken from the excavation sites in cohesive soils for laboratory testing.

The trial pits are generally used for exploration of shallow foundation (depth up to 4 m) . as depths of pits increase, the cost increases very rapidly particularly in water bearing strata. Trial pits are also used to explore depth, cables and other underground public works and /or old structures.

2-Hand and powered augers:

Auger drilling is developing as the most common method of soil exploration to depths up to 60 m. The speed of operation in most types of soil is greater than of any other drilling methods. Distributed sample may be obtained from the soil brought to the surface and undistributed samples may be taken from the bottom of the hole at any required depth. Powered auger with large machines is capable of making borings up to 1.0 m in diameter. In suitable soil where no casing is required, the size of the hole can be examined directly as in trial pits.

2.4.1 Wash borings

In this method, the soil is loosened by a high pressure water jet from a pipe passing down the borehole. The washings are brought to the surface in the water which passes backup the outside of the jet pipe. The methods is extremely cheap and rapid . on the other hand, the washing are usually so disturbed and the sample are used only for identification of soil layers.

2.4.2 Sampling methods:

There are two methods of sampling and they are as follow:

1- Disturbed Samples

There are some instances when sample disturbance is off little important, since only the soil type and stratifications need to be determined. Disturbed sample are more satisfactory where classification of the soil is the prime objective since they are cheaper and faster to obtain (wash boring, auger boring). Disturbed sample are adequate for the investigation of borrow material to be used in fills, highway works and other grading work.

2-Undisturbed samples.

Undisturbed samples are necessary to study soil behavior and to perform laboratory tests under condition as nearly similar to those in the original environment. Shear strength, consolidation and permeability tests must be carried out in the laboratory using undisturbed sample. The major problems associated with undisturbed samples are:

1-The samples are unloaded of their in-situ overburden pressure.

2-Friction of the sites of samplers which creates some disturbance.

3- Samples from levels below G.W.T may drain during the recovery process.

4-Changes in bore pressure will disturb the sample.

Undisturbed soil samples are obtained by forcing a thin walled seamless; stainless steel sampling cylinder into the soil at the bottom of a bore or at the bottom and/ or in the wall of a test pit. The forcing is accomplished by jacking or a continuous push.

In its simplest and more frequently used form, a thin walled sampler consists of a 60 to 90 cm long pipe with the wall thickness of about 1.5 mm and a diameter of 5 to 8 cm. at one end a cutting edge is provided while at other end the type is attached to an adapter filled with a ball value and vents. After samples have been taken, the sampling tube can be detached from the adapter and the ends sealed off with wax for transport to the laboratory. There is may be stored or the wax removed and the sample extruded for immediate testing.

3-Spacing of borings

The spacing of borings cannot be determined with absolute exactness. They depend upon many factors such as the nature and condition of soil , the shape and extent of the structure, etc….farther more the spacing of borings should be conform to the importance, size and system of the structure. The following spacing are often used in planning boring work.

Table 5 soil boring spacing	
Structure or project	Spacing or borings
Highway (sub grade survey)	300-600
Earth dams, dikes	30-60
Borrow pits	30-120
Multistory buildings	15-30
One-story manufacturing plants.	30-90

4- Boring Depth

The depth explored is generally the depth beyond which the effect of structure load is negligible. It should include all stressed zones of soil involved in the foundation-system and should cover all layer of soil which affect the settlement of the structure. The depth of boring may be taken as 1.5 times the breadth of the raft or from 3-5 times the breadth of footing. If sand layer is reached it must be penetrated sufficiently to ensure its continuity.

A better simple rule for structure such as hospital and office building, related the estimated boring depth, d, to the number of stories.

Light steel or narrow concrete $d = 3(S)^{0.7}$ m   (4)

Heavy steel or wide concrete $d = 6(S)^{0.7}$ m   (5)

Where, s, is the number of stories.

5- Boring Report

A boring report should contain:

1- situation plan of construction site drawn to scale and oriented with respect to north.

2- identification, data, place, boreholes location and there coordinates from a reference axis and drilling equipment used.

3- description of soil profiles over the full depth of borehole with classification of soil encountered in each boring. The boring log should be drawn to a stated scale and contain information of soil types and thickness encountered.

4- Description of technique for advancing and stabilizing borehole.

5- location of the ground water table, the ground water condition and/or surface drainage observation.

6- Information on any difficulties or obstruction encountered in boring operations (boulders, roots, drains, telephone and electric cables pipes, old foundation, etc.)

7- listing of samples obtained and report on any testing of soil in place.

2.5 Standard Penetration Test (SPT)

The SPT is a well-established and unsophisticated method, which was developed in the United States around 1925. It has since undergone refinements with respect to equipment

and testing procedure. The testing procedure varies in different parts of the world.

Therefore, standardization of SPT was essential in order to facilitate the comparison of results from different investigations. The equipment is simple, relatively inexpensive and rugged. Another advantage is that representative but disturbed soil samples are obtained.

The reliability of the method and the accuracy of the result depend largely on the experience and care of the engineer on site .

A split-barrel sampler is driven from the bottom of a pre-bored hole into the soil by means of a 63.5 kg hammer, dropped freely from a height of 0.76 m. The diameter of the pre-bored hole varies normally between 60 and 200 mm. If the hole does not stay open by itself, casing or drilling mud should be used. The sampler is first driven to a depth of 15 cm below the bottom of the pre-bored hole, then the number of blows required to drive the sampler another 30 cm into the soil, the so called N30 count, is recorded. The rods used for

driving the sampler should have sufficient stiffness. Normally, when sampling is carried out to depths greater than around 15 m, 54 mm rods are used.

The quality of test results depends on several factors, such as actual energy delivered to the head of the drill rod, the dynamic properties (impedance) of the drill rod, the method of drilling and borehole stabilization. The actually delivered energy can vary between 50 - 80% of the theoretical free-fall energy. Therefore, correction factors for rod energy (60 %) are commonly used, Seed and De Alba (1986). The SPT can be difficult to perform in loose sands and silts below the ground water level (typical for land reclamation projects), as the

borehole can collapse and disturb the soil to be tested. The following factors can affect the test results: nature of the drilling fluid in the borehole, diameter of the borehole, the configuration of the sampling spoon and the frequency of delivery of the hammer blows.

Therefore, it should be noted that drilling and stabilization of the borehole must be carried out with care. The measured N-value (blows/0.3 m) is the so-called standard penetration resistance of the soil. The penetration resistance is influenced by the stress conditions at the depth of the test. Peck et al. (1974) proposed, based on settlement observations of footings, the following relationship for correction of confinement pressure. The measured N-value is to be multiplied by a correction factor CN to obtain a reference value, N1, corresponding to an effective overburden stress of 1 t/ft² (approximately 107 kPa)

$N_1 = N \cdot CN$
(6)

where CN is a stress correction factor and p' is the effective vertical overburden pressure.

$CN = 0.77 \cdot \log 10 (20/p)$
(7)

Figure 11 Rig for SPT

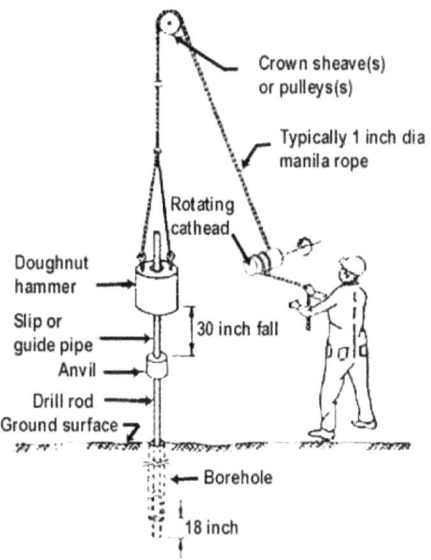

Figure 12 SPT tools

Table 6 Type of the soil and SPT Reading			
Sandy Soil		Clayey or Silty Soil	
N-SPT Value	Relative Density	N-SPT Value	Consistency
0 – 4	Very loose	0 – 2	Very soft
4 – 10	Loose	2 – 4	Soft
10 – 30	Medium	4 – 8	Medium stiff
30 – 50	Dense	8 – 15	Stiff
> 50	Very dense	15 – 30	Very stiff
		> 30	Hard

Table 7 Relation between N60 and su

Description	N60	Su (kPa)
Very soft	0 – 2	<10
Soft	3 – 5	10-25
Medium	6 – 9	25-50
Stiff	10 – 15	50-100
Very stiff	15 – 30	100-200
Hard	> 30	>200

2.6 Cone Penetration Tests (CPT)

The CPT was invented and developed in Europe but has gained increasing importance in other parts of the world, especially in connection with soil compaction projects. Different types of mechanical and electric cone penetrometers exist but the electric cone is most widely used. A steel rod with a conical tip (apex angle of 60° and a diameter of 35.7 mm) is pushed at a rate of 2 cm/s into the soil. The steel rod has the same diameter as the cone.

The penetration resistance at the tip and along a section of the shaft (friction sleeve) is measured. The friction sleeve is located immediately above the cone and has a surface area of 15000 mm². The electric CPT is provided with transducers to record the cone resistance and the local friction sleeve.

A CPT probe, equipped with a pore water pressure sensor is called CPTU. It is important to assure complete saturation of the filter ring of the pore water (piezo) element. Otherwise, the response of the piezo-transducer, which registers the variation of pore water pressure during penetration, will be slow and may give erroneous results. The CPTU offers the possibility to determine hydraulic soil properties (such as hydraulic conductivity - permeability) but is most widely used for identification of soil type and soil stratification.

The CPT can also be equipped with other types of sensors, for example vibration sensors (accelerometer or geophone) for determination of vibration acceleration or velocity. The "seismic cone" is not yet used on a routine basis but has, because of the relative simplicity of the test, potential for wider application especially on soil compaction projects.

The CPT is standardized and the measurements are less operator-dependent than the SPT, thus giving more reproducible results.

The CPT measures the cone resistance (q_c) and the sleeve friction (f_s) from which the friction ratio, FR can be determined. FR is the ratio between the local sleeve friction and the cone resistance, expressed in percent (f_s/q_c). In spite of the limited accuracy of sleeve friction measurements, the valuable information, which can be obtained in connection with compaction projects, has not yet been fully appreciated. As will be discussed below, the sleeve friction measurement reflects the variation of lateral earth pressure in the ground, and can be used to investigate the effect of soil compaction

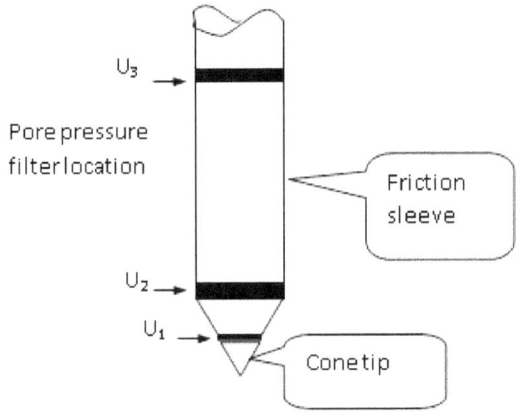

Figure 13 Sketch for CPT

On the state of stress, as will be discussed later. Cone and sleeve friction measurements are also strongly affected by the effective overburden pressure. It is necessary to take this effect into account, similar to the SPT.

One important objective of the CPT investigations in connection with soil compaction is to obtain information concerning soil stratification and variation in soil properties both in horizontal and vertical direction. The friction ratio is often used as an indicator of soil type (grain size) and can provide valuable information when evaluating alternative compaction methods .

Figure 14 Onsite tools for CPT

Measurement of the excess pore water pressure with the CPTU can detect layers and seams of fine-grained material (silt and clay). It is also possible to obtain more detailed data information concerning soil permeability and thus soil stratification.

The CPT accuracy of its data at each measuring function is presented in Table 8 to be as a guideline to know the sensitivity of the measured data.

Table 8 Accuracy Classes

Class	Measured parameter	Min. allowable accuracy	Max. length between measurements (mm)
1	Cone resistance	50 kPa or 3%	20
	Sleeve friction	10 kPa or 10%	
	Pore pressure	5 kPa or 2%	
	Inclination	2 degrees	
	Penetration	0.1 m or 1%	
2	Cone resistance	200 kPa or 3%	20
	Sleeve friction	25 kPa or 15%	
	Pore pressure	25kPa or 3%	
	Inclination	2 degrees	

	Penetration	0.2 m or 2%	
3	Cone resistance	400 kPa or 5%	50
	Sleeve friction	50 kPa or 15%	
	Pore pressure	50 kPa or 5%	
	Inclination	5 degrees	
	Penetration	0.2 m or 2%	

Table 9 Guidance for Cone Resistance in Sand

Soil type	ϕ	N_q	q_c (z =10 m) (MPa)	q_c (z = 20 m) (MPa)
Loose sand	30°	18.4	2.8	5.5
Medium dense sand	35°	33.3	5.2	10.5
Very dense sand	40°	64.2	10.5	21.1

The following table 10 present a practical way to define between different soil type.

Table 10 Consistency of Cohesive Soil

Consistency	Unconfined compressive strength (tons/ft²)	Rule-of-thumb test
Very soft	0–0.25	Core (height twice diameter) sags under own weight
Soft	0.25–0.5	Can be pinched in two by pressing between thumb and finger
Firm	0.5–1.0	Can be imprinted easily with finger
Stiff	1.0–2.0	Can be imprinted with considerable pressure from fingers
Very stiff	2.0–4.0	Barely can be imprinted by pressure from fingers
Hard	>4.0	Cannot be imprinted by fingers

2.7 VANE TEST

The Vane is introduced into the borehole to the depth where the measurement of the undrained shear strength is required. Then it is rotated and the torsion force required to cause shearing is calculated. In the following Figure 15, you can see a manual vane shear. The blade is rotated at a specified rate that should not exceed 0.1degrees per second (practically 1degree every 10 sec). The amount of rotation is specified in the green arrow whereas the red arrow has a device that measures the required Torque. The procedure and the equipment typically should follow the procedures suggested by the ASTM D2573-72.

The shear strength of the material is calculated from the Torque by dividing by a constant K which depends on the dimensions and the shape of the vane. More can be found in ASTM D 2573-72.

The results of the corrected Vane Shear Strength measured in-situ are shown in the following plot. The plot is shear strength (corrected) versus time and the test was performed in a rate of 1degree per 10sec. You can see in the plot the peak shear strength and the residual. After the completion of the test the vane was rotated twice and then the residual strength was measured.

Figure 15 Vane test

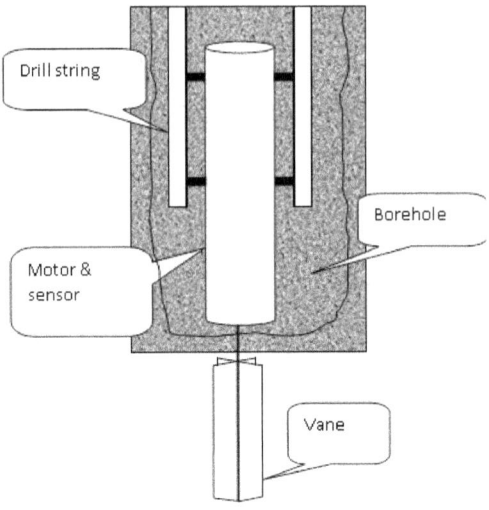

Figure 16 Sketch for van test tool

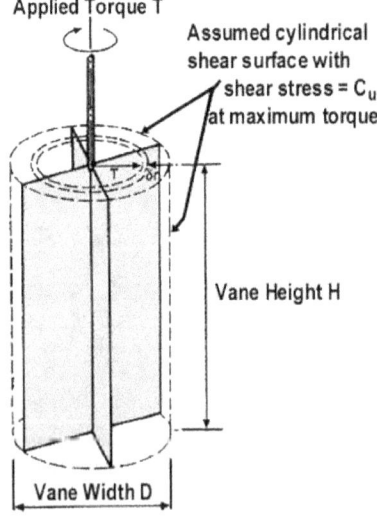

Figure 17 Vane test blade

Table 11 Measuring range with different vane blade dimensions			
Measuring range of s_u	Vane height (mm)	Vane width (mm)	Vane blade thickness (mm)
0–50	130	65	2
30–100	110	55	2
80–250	80	40	2

2.8 Cross hole Test

In case of very critical project as for new facilities with a large storage tanks the phenomena of the soil through the long depth is important in taking an engineering decision.

This test is important to define the wave propagation through the soil depth which is very critical information especially in case of earthquake regions.

The target from this test is to estimate the profile of the seismic velocity of longitudinal and transversal waves along boreholes at the selected location using the cross-hole seismic technique. This test consists of two boreholes 95 m deep drilled using a rotary drilling. The borehole will be cased with 97.6mm diameter PVC pipes to accommodate the source and the receiver probes. The boreholes were spaced around 5 m from center to center at the ground surface.

The inclination of the two boreholes will be measured by using an inclinometer setup. The coordinates obtained from the inclinometer readings were then used to accurately calculate the distance between the two boreholes at 1.0 m depth intervals starting at the ground surface and down to the end of the boreholes.

The raw data which will obtain from the cross hole survey are the travel times of the compression ,P, and shear ,S, waves from the source hole, T, and the receiver hole, R, as shown in the following Figure 18. The travel times of the seismic waves are derived from the fist arrivals identified on the seismic trace for each T and R position with the known distance to calculate the apparent velocities P and S for each depth interval.

This techniques is used in case of special structures and in case of power generation projects for huge power turbine or any vibrating machine.

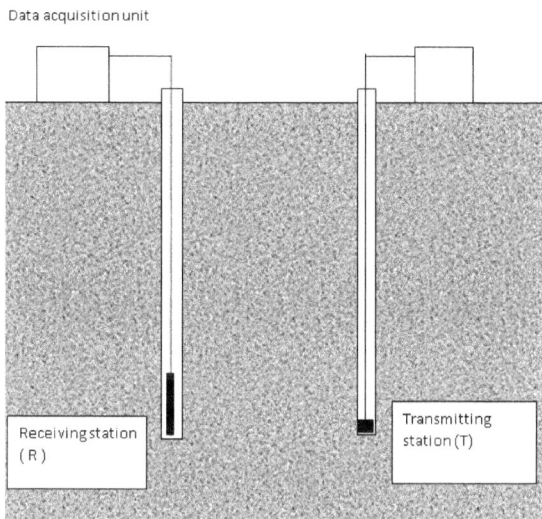

Figure 18 Schematics of cross hole test setup

The instrumentation and data acquisition consists of a probe , a control unit and a cable with cable reel. The probe is characterized by a reliable clamping system, obtained through the progressive bending of a harmonic steel spring aside the probe. There will be a continuous current motor, controlled by the surface electronics, moves the piston inside the probe, controlling the bending of the spring and provided for the clamping and unclamping operations. Inside the probe there are three 10Hz geophones, oriented along the x-y-z axis , to determine the arrival time from the seismic waves of type ,S, to the geophone.

A specially designed down hole hammer that fits through the 10 mm casing will be used as the transmitting source . The hammer has gripper plates that expand out to make firm contact with inside of the casing. The hammer operation is pneumatic and electronic.

The equipment for the cross hole test has many design features which facilitate accurate shear wave velocity measurement s in according with the procedures described in ASTM D4428/D4428M-91 for Dross-Hole seismic testing.

The velocity of propagation of stress waves through infinitive media is a function of the material properties of the media, and depends upon the elastic modulus of the material, poison ratio, and the material density.

2.8.1 Body waves

These waves can be primary or P-wave. This is a longitudinal wave in which the directions of motion of the particles are in the direction of propagation. This motion is non rotational and the wave is one of dilation propagated speed.

Body waves can also be secondary or shear or S-waves. This is a transverse wave in which the direction of motion of the particles is perpendicular to the direction of propagation.

Compressive waves are the fastest followed by the shear waves and subsequently followed by various types of waves. A fluid such as pore water does not transmit shear waves. Accordingly, in soil, shear waves are conducted through the soil skeleton only.

2.8.2 Surfaces waves

In a uniform, infinite medium only P and S wave appear. If the medium is bounded or non-uniform as surface soils there are other simple types of wave may appear. The most important are the surface waves that are propagated near the surface of a solid. Surfaces waves have depths of penetration depending on their wavelength. In non-uniform media, surface waves travel at a velocity that depends on their frequency.

2.9 Plate Load Test

Plate Load Test is a field test for determining the ultimate bearing capacity of soil and the likely settlement under a given load. The Plate Load Test basically consists of loading a steel plate placed at the foundation level and recording the settlements corresponding to each load increment. The test load is gradually increased till the plate starts to sink at a rapid rate. The total value of load on the plate in such a stage divided by the area of the steel plate gives the value of the ultimate bearing capacity of soil. The ultimate bearing capacity of soil is divided by suitable factor of safety (which varies from 2 to 3) to arrive at the value of safe bearing capacity of soil.

- Measure strength and deformation of soil
- Use to determine bearing capacity of soil and its settlement especially for shallow foundation
- Work mechanism : push the circle/square plate at the certain depth with load of 2 – 3x design load until rupture
- Loading influence : 1.5 – 2x plate width
- Relationship to undrained shear strength:

$$S_u = (q_u - \gamma_t . H)/N_c \qquad (8)$$

q_u = rupture load

γ_t = unit weight of soil

H = thickness of soil on the sample surface

N_c = bearing capacity factor

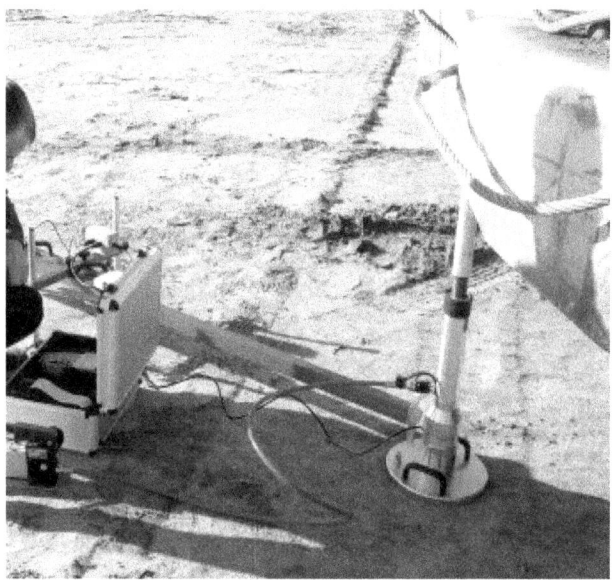

Figure 19 Load test onsite

- The load intensity and settlement observations of the plate load test are plotted in the form of load settlement curves.

- The following figure 20 shows four typical curves applied to different soils.

- **Curve A** is typical for loose to medium non-cohesive soils. It can be seen that initially this curve is a straight line, but as the load increases it flattens out. There is no clear point of shear failure.

- **Curve B** is typical for cohesive soils. This *may* not be quite straight in the initial stages and leans towards settlement axis as the settlement increases.

- **Curve C** is typical for partially cohesive soils.

- **Curve D** is typical for purely dense non-cohesive soil.

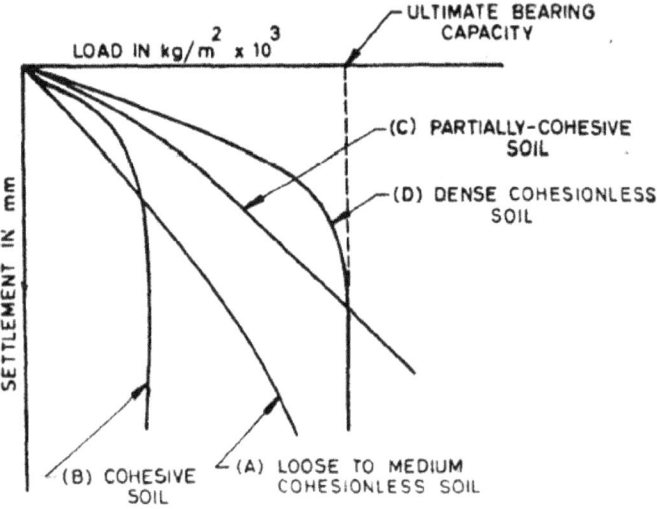

Figure 20 Relation between load and settlement

The safe bearing capacity is obtained by dividing the ultimate bearing capacity by a factor of safety varying from 2 to 3. The value of safe bearing capacity thus arrived at, is considered to be based on criterion of *shear failure.*

Safe bearing capacity (SBC) based on permissible settlement. As indicated earlier the settlement of footing is also related to the SBC of the soil. The value of ultimate bearing capacity and hence the SBC in this case, can be obtained from the load settlement curves by reading the value of load intensity corresponding to the desired settlement of test plate. The value of permissible settlement *(Sf)* for different types of footings (isolated or raft) for different types structures are specified in the I.S. code. The corresponding settlement of test plate (Sp) can be calculated from the following formula,

$$S_f = S_p \left\{ \frac{B(B_p + 0.3)}{B_p(B + 0.3)} \right\}^2 \quad (9)$$

Where,

B = width of footing in mm.

B_p = width of test plate in mm.

S_p = settlement of test plate in mm.

S_f = settlement of footing in mm.

Limitation of plate load test

The plate load test, though very useful in obtaining necessary information about soil for design of foundation has following limitations,

The test results reflect only the character of the soil located within a depth of less than twice the width of bearing plate. Normally the foundations are larger than the test plates, the settlement and shear resistance of soil against shear failure will depend on the properties of much thicken stratum. Thus, the results of test could be misleading if the character of the soil changes at shallow depths.

The Plate Load Test being of short duration, does not give the ultimate settlements particularly in case of cohesive soils.

For clayey soils the bearing capacity (from shear consideration) for a large foundation, is almost same as that for the smaller test plate. But in dense sandy soils the bearing capacity increases with the size of the foundation and hence the test with smaller size test plate tends to give conservative values in dense sandy soils.

In view of the above limitations, the plate load test method of determining SBC of soil may be considered adequate for light or less important structures under normal condition. However, in case of unusual type of soil stratum and for all heavy and important structures, relevant laboratory tests or field test are essential to establish the SBC of soil.

3 Soil investigation report

3.1 Soil investigation SOW

For providing the scope of work for geotechnical soil investigation or reviewing the outcomes from the report the following check in this can assist you in the respective.

The final report of the geotechnical consultant shall include information and recommendations on items marked in the following list, as a minimum (Owner to indicate project / site specific requirements).

1. **Introduction**
   - [ ] 1.1 Description of proposed construction
   - [ ] 1.2 Purpose and scope of investigation
   - [ ] 1.3 Abstract of findings and recommendations

2. **Site Conditions**
   - [ ] 2.1 Site geology, general description
   - [ ] 2.2 Potential geologic hazards
   - [ ] 2.3 Site surface description
   - [ ] 2.4 Site topography, general description
   - [ ] 2.5 Description of aboveground obstructions

3. **Subsurface Conditions**
   - [ ] 3.1 Stratigraphy

- [ ] 3.2 Subsurface material properties, general description
- [ ] 3.3 Groundwater elevations and expected variations
- [ ] 3.4 Description of underground obstructions encountered or otherwise identified

4. Field Investigation

- [ ] 4.1 Summary of operations
- [ ] 4.2 Description of sampling procedures
- [ ] 4.3 Description of field tests
- [ ] 4.4 Logs of borings, soundings, pits, wells, etc., in accordance with *ASTM D5434* and containing the following:

    4.4.1 Complete descriptions and thicknesses of all strata, including near-surface materials such as paving, base course, topsoil, fill, etc.

    4.4.2 Locations referenced to plant coordinate system

    4.4.3 Ground surface elevations referenced to plant datum, if available; if not, then referenced to mean sea level (MSL)

    4.4.4 Standard penetration test values in blows per 150 mm increment

    4.4.5 Results of all field tests

- [ ] 4.5 Location plan, containing as a minimum

    4.5.1 Scale plan with locations of borings, soundings, pits, wells, etc.

    4.5.2 Plant coordinate system

5. Laboratory Tests

- [ ] 5.1 Description of tests
- [ ] 5.2 Test results

6. **Hydrology**
- ☐ 6.1 Erosion potential
- ☐ 6.2 Surface run-off coefficients
- ☐ 6.3 Percolation

7. **Seismic Analysis**
- ☐ 7.1 Seismicity based on seismic risk map
- ☐ 7.2 Soil profile type and site coefficient(s)
- ☐ 7.3 Site specific seismic risk study

8. **Foundation Recommendations**
- ☐ 8.1 Type(s) of foundation recommended
- ☐ 8.2 Basis for selecting recommended foundation type(s)
- ☐ 8.3 Recommendations for foundation type(s) selected
- ☐ 8.4 Recommendations for deep foundations regardless of foundation type selected
- ☐ 8.5 Recommendations for shallow foundations regardless of foundation type selected
- ☐ 8.6 Soil strength parameters used in determining design capacities

9. **Shallow Foundation Recommendations**
- ☐ 9.1 Spread footings: Depth below grade, size, and shape restrictions
- ☐ 9.2 Mat foundations: Depth below grade, modulus of subgrade reaction
- ☐ 9.3 Tank foundations: Recommendations and restrictions, excavation and backfill, ringwall or mat considerations, extended water tests
- ☐ 9.4 Vibratory equipment foundations: Dynamic shear modulus, Poisson's ratio, other considerations

☐   9.4.1   Based on correlations from published literature

☐   9.4.2   Based on in-situ testing

☐   9.5   Ultimate and allowable net soil-bearing capacity

☐   9.5.1   As a function of the shape and size of foundation, depth of embedment, and soil strength

☐   9.5.2   Any increase in net allowable bearing capacity for hydrotest loads, and short-term loads such as wind and earthquake

☐   9.6   Foundation settlement

          9.6.1   As a function of loading, shape and size of foundations, and compressibility of sub-soils

          9.6.2   Immediate settlement during construction

          9.6.3   Long-term settlement

          9.6.4   Time rate of settlement

          9.6.5   Adjacent foundation settlement

          9.6.6   Differential settlement for tanks

                  9.6.6.1   Along the perimeter

                  9.6.6.2   Center of tank to perimeter

                  9.6.6.3   Slope of tank bottom after anticipated settlement

                  9.6.6.4   Limitations or recommendations for hydrotest procedures to minimize differential settlement

                  9.6.6.5   Anticipated settlement and rebound during hydrotest and specific measurements during hydrotest

10. Deep Foundation Recommendations

- [ ] 10.1 Type of pile or drilled pier and basis for recommendation
- [ ] 10.2 Ultimate and allowable axial compression capacity through end bearing and skin friction
    - 10.2.1 Capacity vs length
    - 10.2.2 Any increase in capacity for hydrotest loads, or for short term loads such as wind and earthquake
- [ ] 10.3 Minimum and maximum tip elevations, when applicable
- [ ] 10.4 Ultimate and allowable axial uplift capacity
    - 10.4.1 Uplift capacity vs. length
    - 10.4.2 Any increase in capacity for hydrotest loads, or for short term loads such as wind and earthquake
  - 10.5 Allowable lateral capacity
    - [ ] 10.5.1 Applied lateral loading vs. deflection of pile head
    - [ ] 10.5.2 Pile moment vs. depth
    - [ ] 10.5.3 P-Y curves
    - [ ] 10.5.4 Recommendations for generation of P-Y curves and required parameters (cohesion, friction angle, E_{50})
- [ ] 10.6 Down drag considerations
- [ ] 10.7 Spacing, group action, and use of batter piles
- [ ] 10.8 Settlement considerations
- [ ] 10.9 Vibratory equipment foundations, spring constants in each direction for recommended pile type
- [ ] 10.10 Driven pile installation considerations
    - [ ] 10.10.1 Driving criteria, including refusal criteria

- [ ] 10.10.2 Wave equation analysis
- [ ] 10.10.3 Pre-drilling requirements/restrictions
- [ ] 10.10.4 Potential problems and recommended solutions
- [ ] 10.10.5 Pile installation near existing facilities
- [ ] 10.11 (Non-driven) drilled pile and pier installation considerations
    - 10.11.1 Installation equipment requirements
    - 10.11.2 Casing/slurry considerations
    - 10.11.3 Installation criteria and recommendations
    - 10.11.4 Potential problems and recommended solutions
- [ ] 10.12 Load test requirements, procedures, and acceptance criteria

11. Earth Pressures

- [ ] 11.1 Active earth pressure, at-rest earth pressure
- [ ] 11.2 Ultimate and allowable passive soil resistance for on-site soils, and recommended fill and backfill material
- [ ] 11.3 Groundwater considerations
- [ ] 11.4 Drainage requirements

12. Soil Properties

- [ ] 12.1 Coefficient of friction or adhesion values between soil and concrete
- [ ] 12.2 Unit weight of soil
- [ ] 12.3 Cohesion and angle of internal friction
- 12.4 Chemical analysis and other properties of soil and groundwater at depths of proposed structural elements and utilities, as follows:
    - [ ] 12.4.1 pH value

- [ ] 12.4.2 Electrical conductivity (laboratory determination)
- [ ] 12.4.3 Chloride ion (Cl) concentration
- [ ] 12.4.4 Sulfate ion (SO$_4$) concentration
- [ ] 12.4.5 Electrical resistivity of soil (field determination)
- [ ] 12.4.6 Yearly average moisture content of soil
- [ ] 12.4.7 Thermal resistivity of soil

12.5 permeability
- [ ] 12.5.1 Laboratory determination
- [ ] 12.5.2 In-situ determination

13. Slabs and Pavements

- [ ] 13.1 Natural soil and fill, subgrade suitability
- 13.2 Recommended California bearing ratio value for pavement design
  - [ ] 13.2.1 Based on correlations from published literature
  - [ ] 13.2.2 Based on laboratory testing
  - [ ] 13.2.3 Based on in-situ testing
- [ ] 13.3 Recommended modulus of subgrade reaction for slab design
- [ ] 13.4 Treatment for improving subgrade, if required
- [ ] 13.5 Base course, sub-base course, and shoulder recommendations
- [ ] 13.6 Surfacing recommendations
- [ ] 13.7 Base, sub-base, and subgrade drainage recommendations
- [ ] 13.8 Complete pavement system design

14. Other Considerations; Discuss and Provide Recommendations for the Following:

- [ ] 14.1 Frost susceptibility of soils, frost depth
- [ ] 14.2 Liquefaction potential of soils
- [ ] 14.3 Swelling potential of soils, including depth of zone of soil moisture content fluctuation
- [ ] 14.4 Collapsible or dispersive soils
- [ ] 14.5 Effects of proposed construction on existing facilities or adjacent property
- [ ] 14.6 Geologic or other potential hazards

15. Excavation Considerations

- [ ] 15.1 Allowable excavation slope inclinations, temporary and permanent
- 15.2 Groundwater control
  - [ ] 15.2.1 Recommended dewatering method
  - [ ] 15.2.2 Temporary and permanent groundwater control
  - [ ] 15.2.3 Flow quantities
- 15.3 Foundation subgrades
  - [ ] 15.3.1 Heave control
  - [ ] 15.3.2 Protection/preserving integrity of subgrade
- [ ] 15.4 Effects on existing facilities
- [ ] 15.5 Potential excavation problems
- 15.6 Rock excavation
  - [ ] 15.6.1 Rippability of rock
  - [ ] 15.6.2 Definition of rock for contract documents
  - [ ] 15.6.3 Rock quantity estimate guidance
- [ ] 15.7 Pressure diagrams for shoring design

- [ ] 15.8 Applicability of specialized shoring/stabilization procedures
- [ ] 15.9 Classification of soil types per OSHA regulations (Types A, B, C)

16. Dikes And Embankments
- [ ] 16.1 Recommended slope inclination
- [ ] 16.2 Slope stability analysis
- [ ] 16.3 Settlement
- [ ] 16.4 Seepage analysis
- [ ] 16.5 Erosion protection of slopes
- [ ] 16.6 Foundation and subgrade preparation
- [ ] 16.7 Fill material: Type, compaction, and moisture content control

17. Railroads
- [ ] 17.1 Natural soil and fill subgrade preparation
- [ ] 17.2 Ballasting, with consideration to availability of local materials

18. Earthwork
- [ ] 18.1 Topsoil: Thickness for stripping; definition for contract documents
- [ ] 18.2 Suitability of on-site material for structural and non-structural fill
- [ ] 18.3 Special preparations or other requirements for use of on-site material
- [ ] 18.4 Availability of imported fill
- [ ] 18.5 Subgrade preparation
- [ ] 18.6 Recommended compaction criteria and moisture content control

☐ 18.7 Potential compaction difficulties and recommended solutions

3.2 Soil investigation report

The soil report should have the main key elements which are as follow:

- Situation plan of construction site drawn to scale and oriented with north direction.
- Identify the boreholes location and their coordinates.
- Soil Bearing capacity
- Description of soil profile (Silty soil or not)
- Define the location of the ground table
- Presence to sulphate
- Recommendation to concrete cover
- Recommendation of the tie beam dimension

4. Foundation Design

4.1 Shallow Foundation

4.1.1 Isolated footing

Isolated or single footings are used to support single columns. This is one of the most economical types of footings and is used when columns are spaced at relatively long distances.

The previous methods calculate the net allowable bearing pressure, so based on that select the dimensions of the foundation and then calculate the thickness of it.

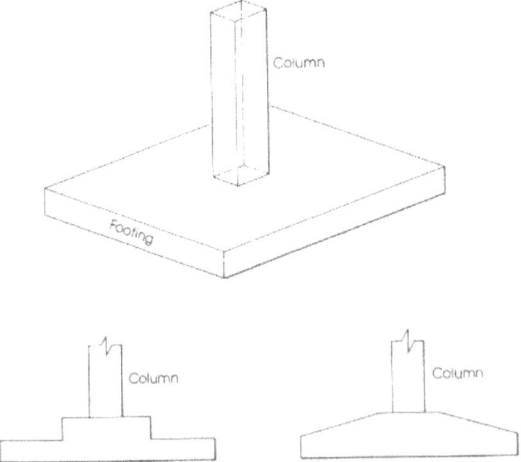

Figure 21 Shapes of shallow foundation

$$q_{all} = P/(L \times B) \qquad (10)$$

Where

q_{all} = allowable bearing pressure

P=is the load from the column   (it will be without factor Dead load+Live load)

L=length of the foundation

B= breath of the foundation

Distribution of Soil Pressure

When the column load P is applied on the centroid of the footing, a uniform pressure is assumed to develop on the soil surface below the footing area.

However the actual distribution of the soil is not uniform, but depends on may factors especially the composition of the soil and degree of flexibility of the footing.

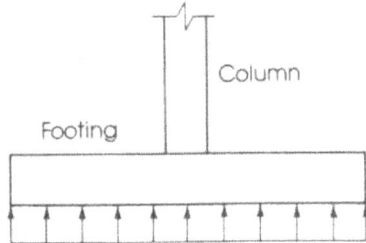

Figure 22 Soil stresses under foundation

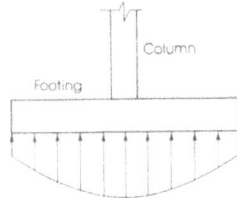

Figure 23 Soil pressure distribution in cohesionless soil

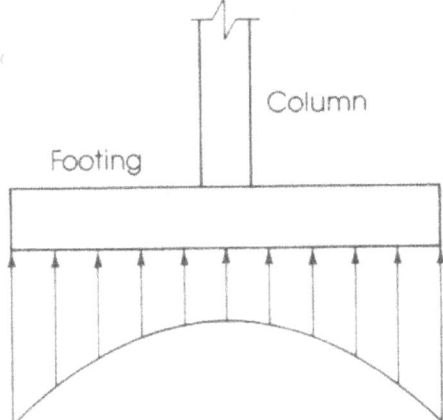

Figure 24 Soil pressure distribution in cohesive soil

Example of this as follow, the excel; sheet is available contact us in www.elreedyman.com to send to you

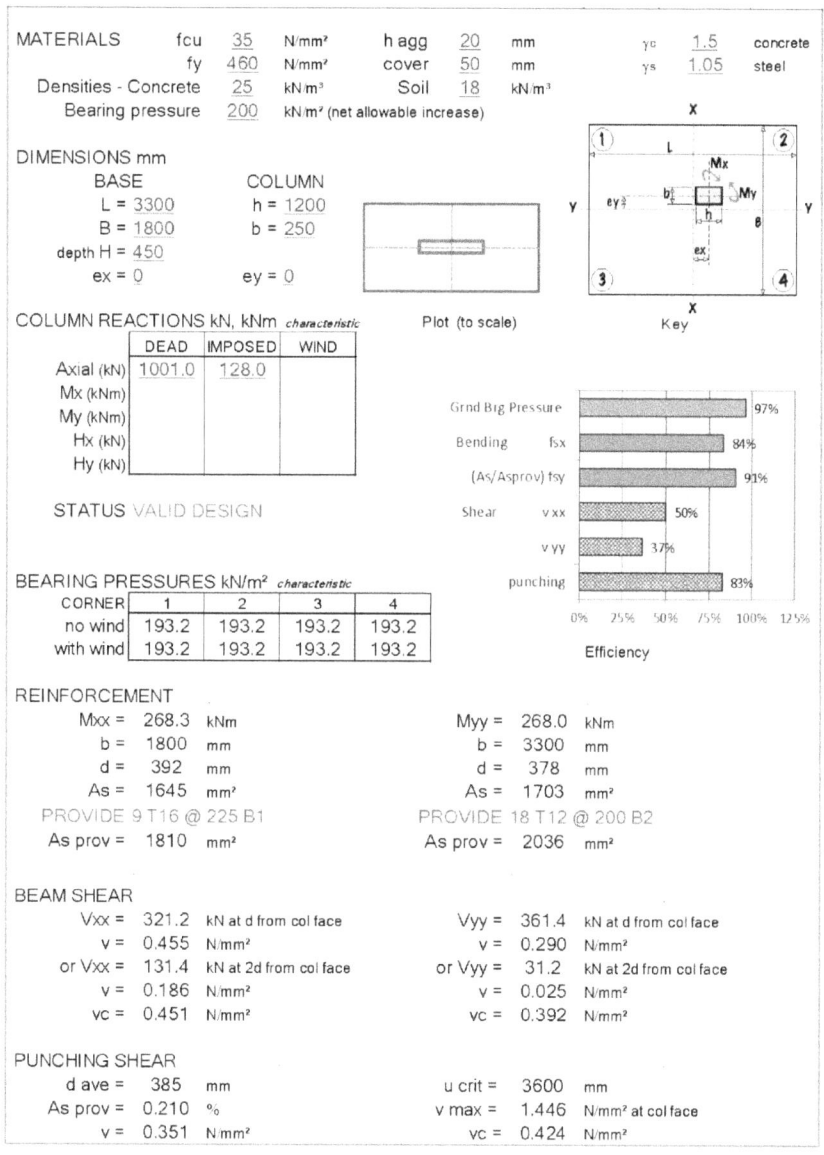

Considering that the column length is, h and width is , b, during selecting the foundation dimensions considering

h/b=L/B
    (11)

4.1.2 Combined footing

In case of the columns is near to each other so the shallow footing will be overlap. In this case it will be converted to combined footing.

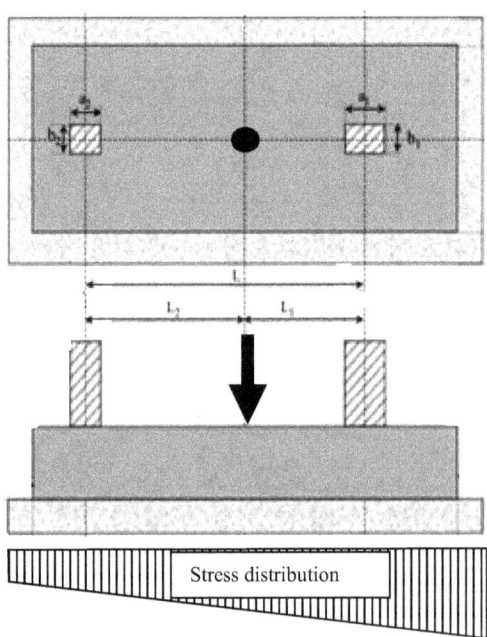

Figure 25 Combined footing stress distribution

The designer will do his best to avoid the combined footing as the steel reinforcement may be reach 250 kg/m3 of concrete however, in case of isolated footing round 100 kg/m3.

As per Figure 22 the resultant force will be toward the direction of the bigger load of the column, based on that it will be an eccentricity on the footing. The design can be same as the isolated footing and sometimes it can be designed like a flat slab by using finite element.

4.1.3 Cantilever or strap footings

It consists of two single footings connected with a beam or a strap and support two single columns. This type replaces a combined footing and is more economical.

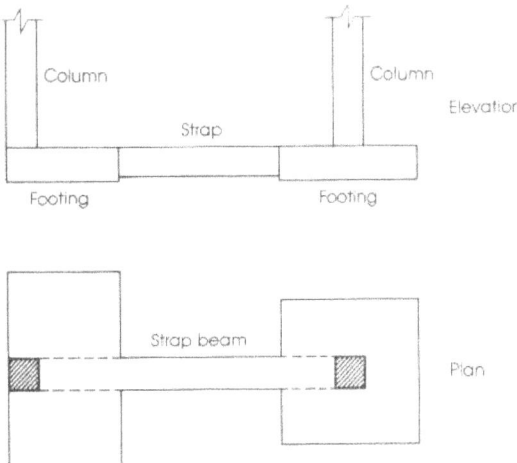

Figure 26 Strap footing

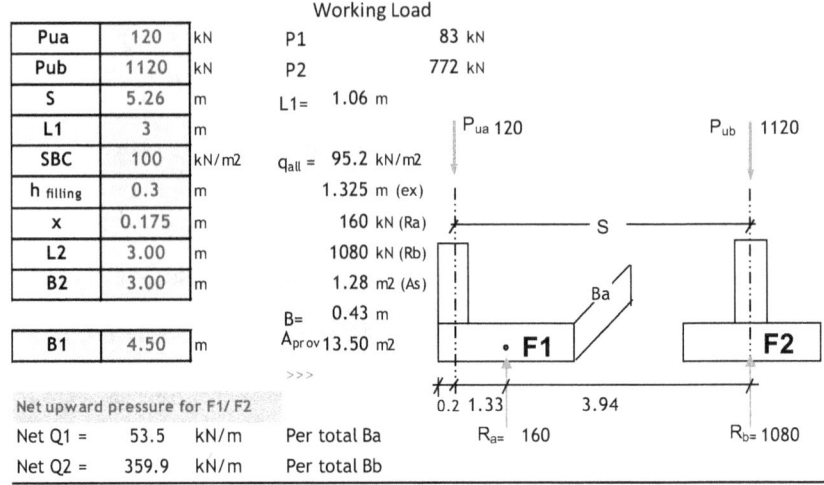

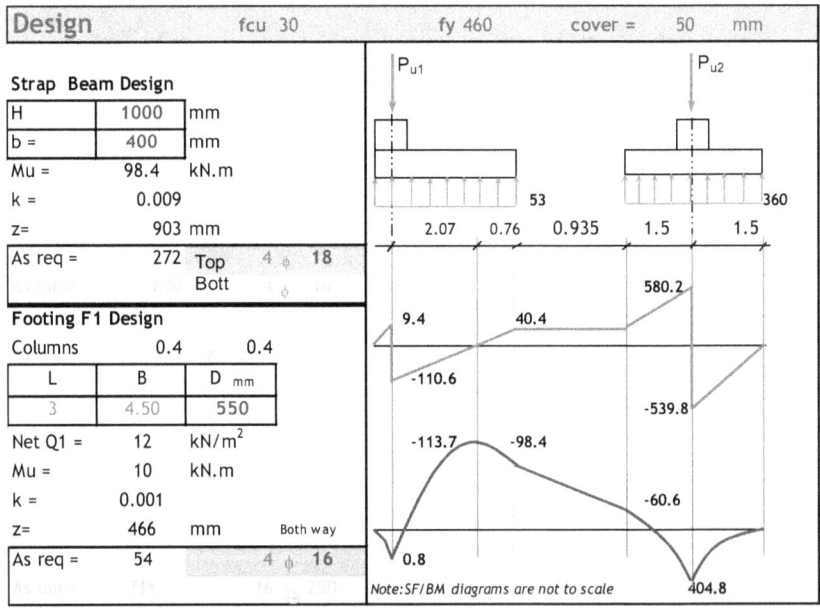

Figure 27 Example for strap footing

The target of the strap footing is to do stability of the end footing to the nearest isolated footing. The bending moment on the strap footing is presented in Figure 24. From this curve the moment on the strap beam is higher on the top so the heavy steel will be on the top.

4.1.4 Raft foundation

In case of increasing the load and this is in case of multi-story building and depend on the soil bearing capacity , the designer shall use the raft foundations as in most cases the distance between the columns about 4 m so in case of foundation have a dimensions over 4 m so it will be an overlap due to this scenario the raft foundation is the suitable solution. In the past the raft foundation design is a slab and beam as in the following figure and the soil pressure consider as the load to the slab and design like a normal slab design.

Now a day this method is not use as it is not accurate and also not economic so I will be one raft with thickness around 1 m depth and the reinforcement placement like the flat slab and the design will be by finite element analysis.

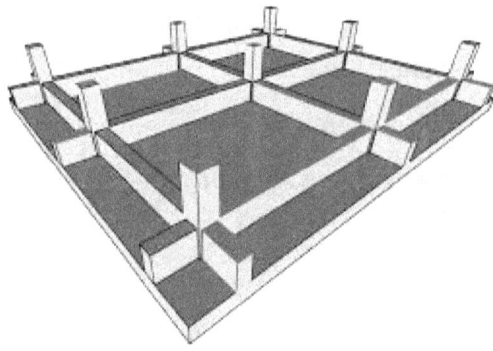

Figure 28 Old approach for raft foundation design

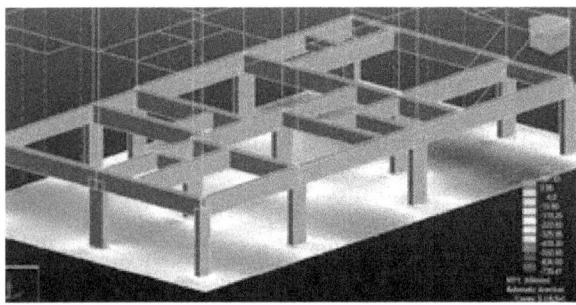

Figure 29 Raft foundation modeling

4.2 Deep Foundation

Pile foundation is the most common type of deep foundations used to transmit the structural loads into the deeper layers of firm soil in such a way that these layers of soil or rock can sustain the loads. A pile foundation, in general, is more expensive than an ordinary shallow foundation, and is used where soil at or near the surface is of poor bearing capacity or settlement problems are anticipated.

The main functions of piles are:

1- Carry more load from the superstructure to the lower, more resistant soil strata, thus increasing the load capacity of the site.

2- To reduce the settlements to the minimum value and consequently the differential settlements. They are most effective in case of sensitive structures which by virtue of their case of sensitive structural statically system cannot undergo appreciable differential settlements.

3-To avoid excavation under water for sites where G.W.T. is high. This may represent an expensive item in the cost of foundation and may also cause reduction of strength of some certain soils.

4- To densify the soil by driving compaction piles in loses cohesionless soil deposits.

Generally, piles are made of timber, steel, concrete, or combined of these materials.

Technically from design point of view, there are two types of pile which are end bearing piles and friction piles. The two options have advantage and disadvantages.

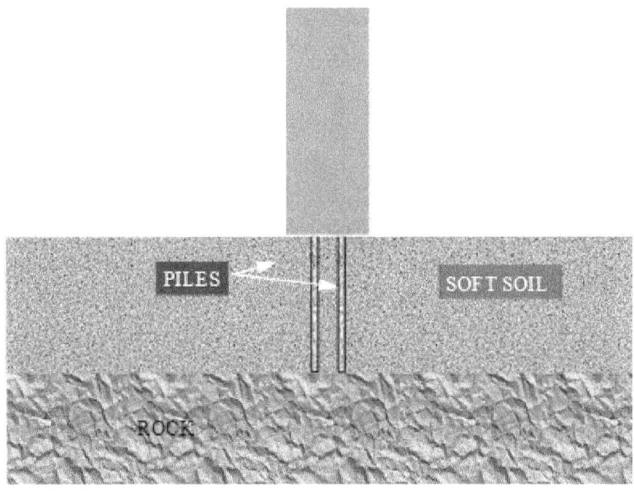

Figure 30 End bearing pile

From Figure 26, it presents the end bearing pile, in this case, most of the load transfer to the rock layer by the end bearing of the pile. So the most loads will be taken by its base and the rest of the load is taken by the section friction.

The skin friction along the system of the pile could be neglected and the bearing capacity of the pile is derived only from the point bearing resistance of the soil under pile tip.

In order to obtain the full benefit of the ultimate strength of firm layer under the pile tip, the pile should penetrate the bearing stratum to a depth at least three times the pile diameter.

Figure 27 present the friction piles that transfer their loads to the surrounding soil by friction developed along their sides.

If the pile penetrates a clay layer, the skin friction is equal to the cohesion, C, of that layer. In case of granular material, the skin friction is proportional to the intensity of earth pressure and can be considered varying linearly with enough accuracy.

Figure31 Friction pile

There are a different types of piles used and the differences between them depends on its material and way of execution.

4.2.1 Timber piles

They are widely used in woody countries in the past and are made of tree trunks of good timber quality and appreciable size (not less than 300 mm in diameter) as shown in Figure (32) &(33). After the timber pile is driven into the ground, the top end should be cut off square, so that the foundation is in contact with soil wood. If a timber pile is subjected to alternate wetting and drying, it should be treated with a wood preservative to increase its useful life. Timber piles can safely carry load between 15-25 tons for usual conditions. They are relatively of low cost and easy to handle.

Figure 32 Timber pile

Figure 33 Wood pile preparation for hammering

4.2.2 Steel piles

These are usually rolled sections of (H) shapes or steel pipes. Wide-flange & I-beams may also be used.

Splices in steel piles are made in the same manner as in steel columns, i.e. by welding, riveting, or bolting. Pipe piles are either welded or seamless-steel pipe which may be driven either open-ended or closed-ended. The pipe piles are usually used on the offshore structure platforms.

4.2.3 Concrete piles

These types of piles consisting of two methods of construction: the cast in place piles and the precast concrete piles.

- *Cast in place piles*

They are formed by making a hole in the ground and filling it with concrete. The may be drilled or formed by driving a shell. The steel piles as shown in Figure 34 and 35 are used to be driven or make the hole and filling concrete inside it.

The steel shell is usually withdrawn during or after pouring concrete and sometimes is left to protect the concrete from mixing with the mud or cement washed away by the ground water.

Figure 34 steel pile with cone tip

Figure 35 Raymond shell pipe

4.2.4 Pre-stressed Pile

Piles are formed to the specified length, cured, and then shipped to the construction site. A primary consideration with precast piles is the handling stress. To take care of handling stresses, some of which are tensile. The piles are reinforced and in some cases prestressed.

Precast R.C .piles may have square or octagonal cross sections. They should be adequately reinforced to with stand driving and handling stresses. Long precast piles should be driven with careful guides o prevent their buckling during driving in the part of the pile in the driving rig. To overcome the driving stresses the lateral steel reinforcements should become closely spaced at top and bottom of the pile to resist the stress wave concentration at the ends of the driven pile.

Bending moment due to handling for relatively short piles (less than 12.0m), pile is usually lifted from one end and the pile is treated as a beam carrying its own weight. For long piles; the piles should be lifted from two, three, or four points at the specified distances indicated in Figure (.14) in order to reduce he bending stresses to minimum. Lifting points should be marked by hooks or bolts that will be removed later.

Special care should be given to the material of concrete piles to remain intact and prevent aggressive soil from attacking the pile. It is necessary in the first place to use siliceous aggregates and rich cement content (350 kg of cement per m^3 of finished concrete). In addition, concrete piles must be protected from the dissolved sulphates or chemicals present in the underground water. If the ratio of sulpher trioxide (So_3) in the soil water increases over 0.03% (300 m.gm/lit.)in stagnant water, or 0.015 in running water, and if in addition the ratio of So_3 in the soil itself is 0.2%, ordinary Portland cement cannot be used. In this case, special

sulphate resisting cements should be used. In all cement used, the presence of free lime or calcium traces should be minimizing as we can.

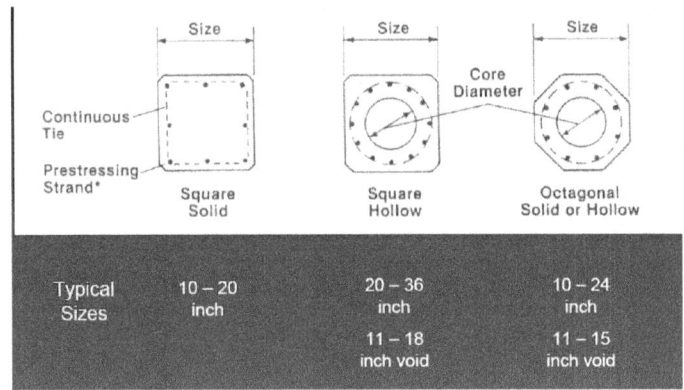

Figure 36 Prestressed pile sections

Figure 37 Lifting for concrete pile

Figure 38 Pile with composite section

The following table (12) is summarize the comparison between different types of piles.

4.2.5 Pile calculation

There are many methods and theory to calculate the pile capacity but the basic concept is shown in Figure which is based on that the bearing capacity (Q_d) of a pile may be assessed using soil mechanics principles. The capacity is assumed to be the sum of skin friction and end-bearing resistance, i.e

$$Q_d = Q_f + Q_p = fA_s + qA_p \quad \ldots\ldots\ldots\ldots\ldots\ldots\ldots\ldots(12)$$

Where

Q_f = skin friction resistance,

Q_p = total end bearing resistance

Table 12 Typical pile characteristic and use

Characteristic	Pile type			
	Concrete filled steel pipe piles	Composite piles	Precast concrete (including prestress)	Cast in place (thin shell driven with mandrel)
Maximum length	Practically unlimited	55m	30m for precast, 60 m for prestress	30 m for straight section, 12 m for tapered section
Optimum length	12-36 m	18-36 m	12-15 m precast 18-30 m prestress	12-18m for straight 5-12 m for tapered
Application material specifications	ASTM A36 for core ASTM A252 for pipe ACI 318 for concrete	ASTM A36 for core ASTM A252 for pipe ACI 318 for concrete ASTM D25 for timber	ASTM A15 reinforcing steel. ASTM A82 cold drawn wire ACI 318 for concrete	ACI
Recommended maximum stress	$0.4f_y$ reinforcement, $0.5f_y$ or core $0.33f_c'$ for concrete	Same	$0.33f_c'$ $0.4f_y$ for reinforcement unless prestress	$0.4f_y$ if steel gauge ≤ 14, $0.35f_y$ if shell thickness $\leq$ 3mm
Maximum load for usual condition	1800 KN with cores 18000KN for large section with steel core	1800 KN	8500 KN for prestress 900kN for precast	675KN
Optimum load range	700-1100 with cores 4500-14000 KN with cores	250-725 KN	350-3500 KN	250-550 KN

F = unit skin friction capacity
A_s = side surface area of pile
q = unit end bearing capacity
A_p = gross end area of pile

There are many theory to calculate the pile capacity for cohesive and non cohesive soil one of this method from API presented below as a guideline for the way of calculating the pile capacity

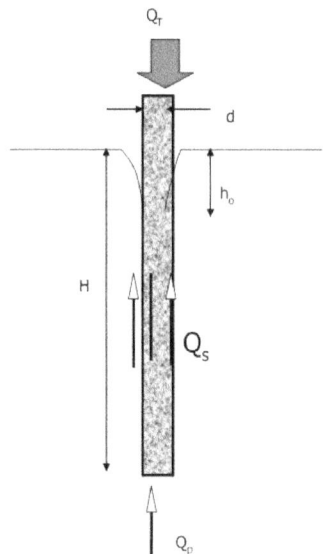

Figure 39 Pile forces

A. For cohesive soil

1. Skin friction

$$f = \alpha \, S_u \qquad (13)$$

α = a dimensionless factor

S_u = undrained shear strength of the soil at the point in question

$$\alpha = 1.0 \ (S_u \leq 25 \text{ kPa}) \qquad (14)$$
$$\alpha = 0.5 \ (S_u \geq 70 \text{kPa}) \qquad (15)$$
$$\alpha = 1 - (S_u - 25/90) \qquad (16)$$

Where P_o is the effective overburden pressure at the point in question

$$N_c = 9 \quad \text{for } L/D \geq 3; \ (s_u)_b > 25 \text{ kPa} \qquad (17)$$
$$N_c = 6 \quad \text{for } (s_u)_b \leq 25 \text{ kPa} \qquad (18)$$

2. End bearing

$$q = 9 \, S_u \qquad (19)$$

B. Non cohesive soil

1. **Skin friction**

$$f = k \, p_o \tan\phi \qquad (20)$$

k = coefficient of lateral earth pressure (ratio of horizontal to vertical normal effective stress)

ϕ = friction angle between the soil and pile wall.

2. **End bearing**

$q = p_o N_q$ (21)

N_q = dimensionless bearing capacity factor

Example:

Calculate the pile capacity for a precast reinforced concrete pile 0.25 x 0.25 m with length 10m it consists of two layer a medium and stiff clay.

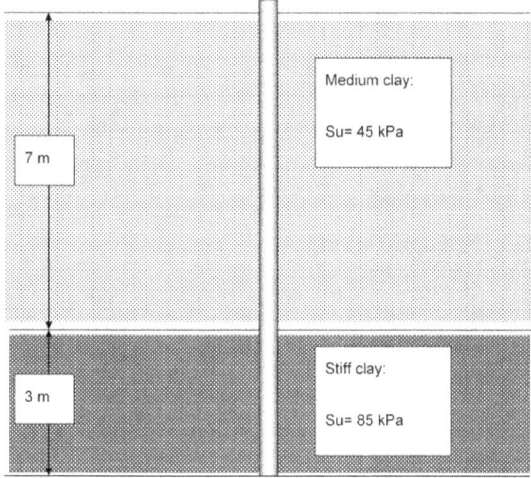

Figure 40 Pile calculation example

Solution:

For medium clay layer:

S_u=45 kPa α= 1-(45-25/90)=0.78

For stiff clay layer

S_u=85 kPa α= 0.5

Friction capacity, Q_{fs}= 0.78 x45x (4 x0.25)x7+0.5 x85x (4 x0.25)x3=373.2 kN

End bearing capacity $Q_b = 9\ S_u\ A_b = 9 \times 85 \times (0.25 \times 0.25) = 47.82$ kN

The pile capacity = Qt/FS = (373.2+47.82)/3 = 140 kN

4.2.6 Pile caps

Piles caps are kind of foundation, which affects by the column loads from up and Piles reaction at the point of contact piles to the caps. In this type of foundation ignore the impact of the soil where the soils are not contact with the caps in rigid or flexible manner to allow to carry any part of the column load and the stiffness of the piles is very heavy so the piles carry all he load.

Often the column load cannot carry by one pile only so the column needs more than one pile to carry the load so it needs a piles cap to distribute the column load to these piles equally.

In phase of design, should care to distribute the load to the piles equally and this can happen by putting the center of gravity of the column coincide with the center o gravity of the pile cap.

To ensure transfer the load from the column to the pile the pile steel reinforcement should extend inside the piles cap with at least 600 to guarantee transferee this load by bond between concrete and steel.

The pile caps will be design as a rigid foundation and the piles carrying equal loads from the column load so the pile cap thickness should resist the punching stresses and the tension in top and in bottom.

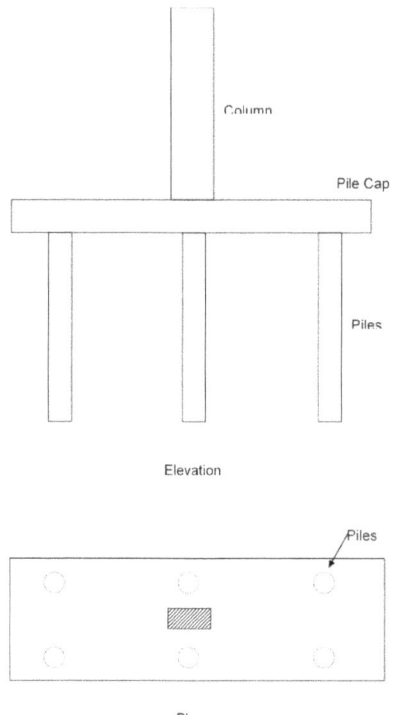

Figure 41 Pile cap

4.2.7 Pile Load test

The purposes of a pile load test are:

- To determine the load capacity of a single pile or a pile group, especially when the design requires methods that are outside of accepted practice.

- To determine the settlement of a single pile at working loads.

- To verify estimated load capacity.

- To obtain information on load transfer in skin friction and in end bearing.

- To satisfy regulatory agencies.

The ASTM D 1143 provides the standard methods for conducting pile load tests. Only a brief description is presented in this section. In a typical pile load test (conventional load test), the test pile is driven to the desired depth, loads are applied incrementally, and the settlement of the pile is recorded.

The axial loads can be applied by stacking sandbags on a loading frame attached to the pile or, more popularly, by jacking against a reaction beam and reaction piles Figure 42. If load transfer information is required, the pile must be instrumented with sensors or strain gages to deduce the strains and stresses along the pile length. The conventional pile load test gives the combined skin friction and end bearing resistance. They cannot be separated easily.

The interpretation of the load capacity depends on the method of loading. Two loading methods are popular.

Figure 42 Pile test

In one method, called the constant rate of penetration (CRP) test, the load is applied at a constant rate of penetration of 1.25 mm/min in fi ne-grained soils and 0.75 mm/min to 2.5 mm/min in coarse-grained soils. In the other method, called the quick maintained load (QML), increments of load, about 10% to 15% of the expected design load, are applied at intervals of about 2.5 min. At the end of each load increment, the load and settlement are recorded. Usually, the maximum load applied is about twice the expected design load for single piles and about one and a half times the expected design load for group piles.

5 Retaining wall

5.1 Introduction

Retaining wall is used to confine water or soil. In industrial project is usually used around the tanks to confine the oil from seepage in case of tank failure or leak of any of the tanks.

In the roads it is used to cover the problem of different level so it needs a retaining wall to overcome these levels.

In general, the following is the types of the retaining wall:

1. Gravity retaining wall
2. Cantilever walls
3. Counter fort walls
4. Buttresses walls
5. Bridge abutments
6. Basement walls
7. Semi gravity walls
8. Walls with relieving slabs

9. Precast and prestressed walls

In process plant, it is traditionally to use the cantilever walls. The forces which effect on the retaining walls after defining the preliminary dimensions to the walls will be as follow:

1. The wall height
2. The dead and live load on the wall
3. The lateral pressure that affect the wall due to the soil weight behind the wall.
4. The lateral pressure due to live load or moving load on the soil that exist behind the wall.
5. The wave effect in case of wall adjacent to the sea.
6. Earthquake load
7. The water or oil lateral pressure in case of bundle wall.
8. Any other load that affect the structure during execution or operation phase.

5.2 Preliminary retaining wall dimensions

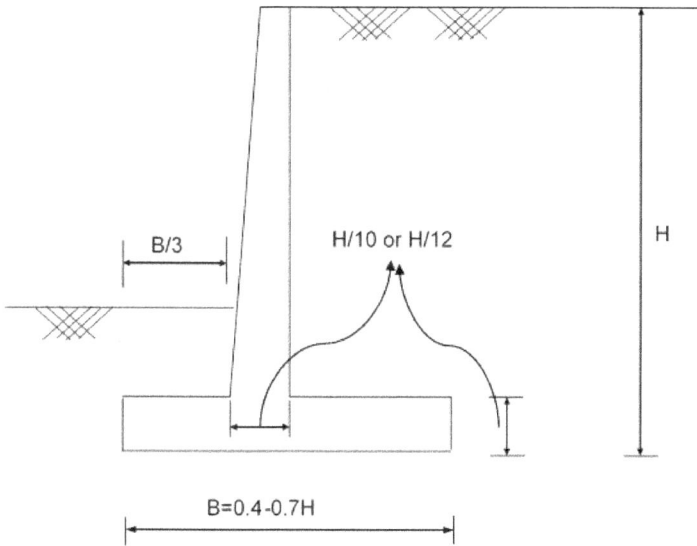

Figure 43 Retaining wall preliminary dimension

After defining the preliminary dimensions as shown in the Figure 43, the design of the retaining wall will start by check the stability of the retaining wall against the following three criteria:

1. Stability against overturning
2. Stability against sliding
3. Stability against bearing capacity

After check the stability of the retaining wall then calculate the stress on the wall to design the reinforced concrete section of the wall and check stability again for second round until the compatibility check of stability with safe reinforced concrete design sections dimensions

5.3 Check stability against overturning

To check stability of the retaining wall against overturning will be done through applying the following equation which is the sum of moments due to forces tending to resist overturning should be higher than the sum of the moments due to forces tending to overturn the walls.

$$FS = \frac{\sum M_R}{\sum M_O}$$
(22)

Where,

MR= Some of the moments due to forces tending to resist overturning about point (C) as in Figure 44.

MO= Some of the moments due to forces tending to overturning the wall about point (C) as in Figure 44.

FS= factor of safety about 1.5-2

The following figure presents the moments that affect the wall overturning and resistance against overturning.

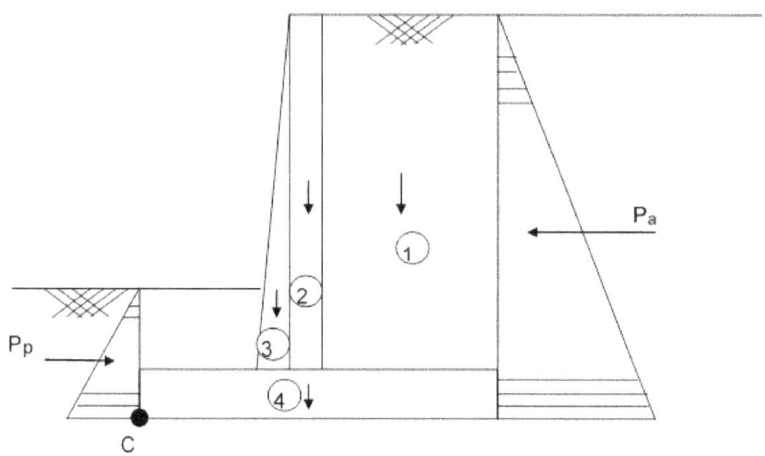

Figure 44 Forces produce and resist overturning

$P_a = K_a \gamma H$ (23)

Where Pa is the active force and Ka is equal to 0.3.

$$\sum M_o = P_a(H/3)$$ (24)

Where, MR can be calculated from the following table 13.

γ_c = concrete density

γ_s = soil density

Calculate FS if within 1.5-2.0 accept if less it needs to change the retaining wall dimensions to achieve the required overturning stability factor of safety.

Table 13 resisting overturning moment				
Section	Area	weight	Moment arm measure from (C)	Moment about (C)
1	A1	W1=A1 x γs	X1	M1
2	A2	W2=A2 x γc	X2	M2
3	A3	W3=A3 x γc	X3	M3
4	A4	W4=A4 x γc	X4	M4
		ΣV		ΣMR

5.4 Check stability against sliding

$$FS = \frac{\sum F_R}{\sum F_d}$$
(25)

Where:

ΣF_R = sum of the forces that resist the retaining wall from sliding

ΣF_d=Sum of the horizontal driving forces to slide the retaining wall

$\Sigma F_d = P_a$

$\Sigma F_R = SV \times \tan\phi 2 + PP$

$= SV \tan(k1\phi^2) + k_2 C_2$

(26)

Where, $k_1 = k_2 = 2/3$

P_P is the passive forces but from a practical point of view it shouldn't take into calculation to increase the factor of safety. However usually during construction filling in front of the wall will be the last stage.

If the factor of safety is less than the limit, you can increase the stability against sliding by using a key lock as shown in the following factor.

5.4 Check stability against bearing capacity

The soil bearing capacity is obtained from the soil report. The retaining wall should check that the load which transfer from the wall to the soil less than the soil bearing capacity.

$$FS = \frac{q_u}{q_{max}}$$

(27)

Where,

q_{all}= maximum allowable bearing capacity

q_{max}=maximum applied load to the soil.

FS= factor of safety between 2.5 -3

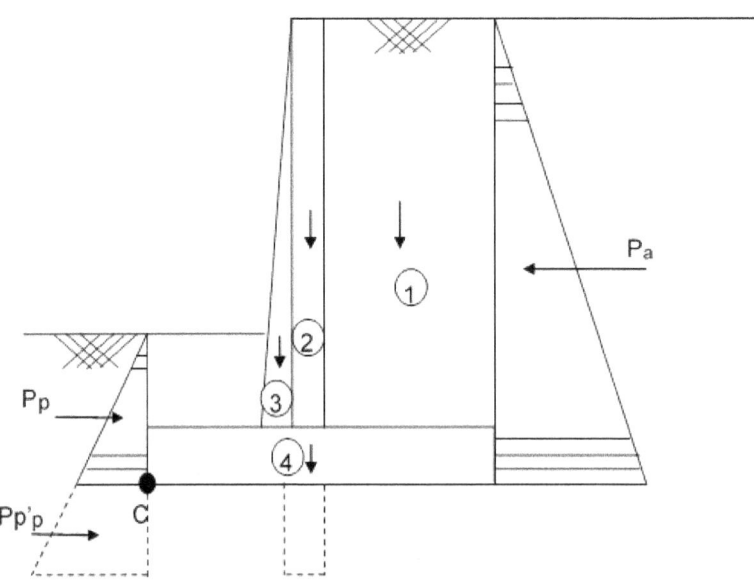

Figure 45 Forces drive and resist sliding

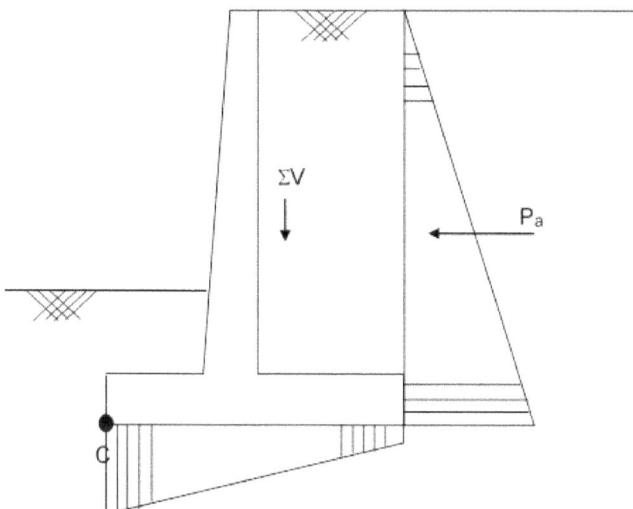

Figure 46 Stresses check against bearing capacity

$M_n = \Sigma M_R - \Sigma M_o$

(28)

$X = M_n / \Sigma V$

$e = B/2 - X$

$$q_{min} = \frac{\Sigma V}{B}\left(1 - \frac{6e}{B}\right)$$
(29)

$$q_{max} = \frac{\Sigma V}{B}\left(1 + \frac{6e}{B}\right)$$
(30)

The second step is to start design the concrete sections. In the following Figure 47 the main forces that affect each element is presented.

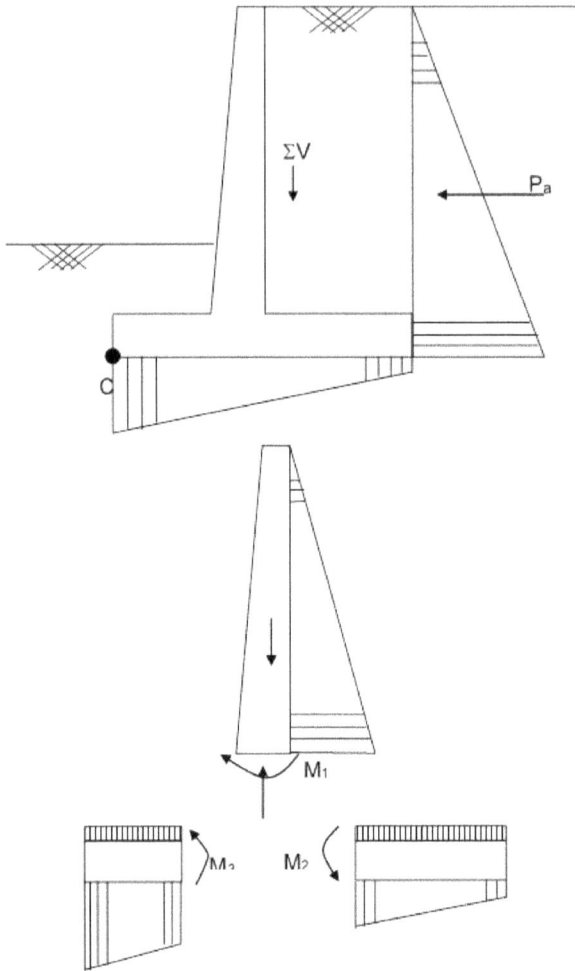

Figure 47 Stresses check against bearing capacity

$M_1 = M_2 + M_3$
$$(31)$$

The shape of the steel reinforcement detail will be as shown in Figure.48 and the steel bending detail for the steel bars embedded in the foundation and extend in the stem is presented in Figure 49.

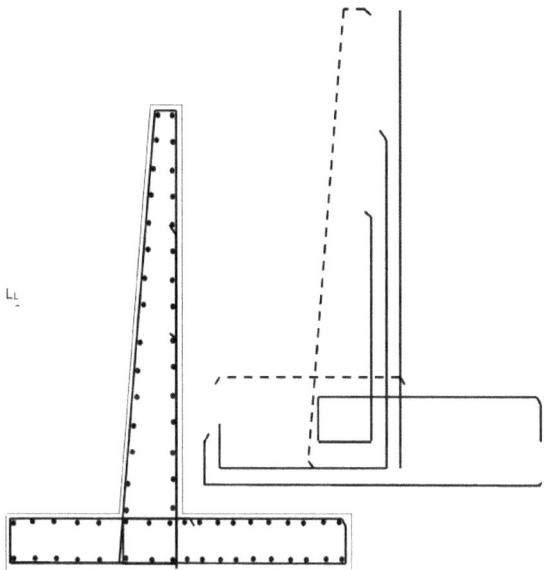

Figure 48 Retaining Wall steel detailing

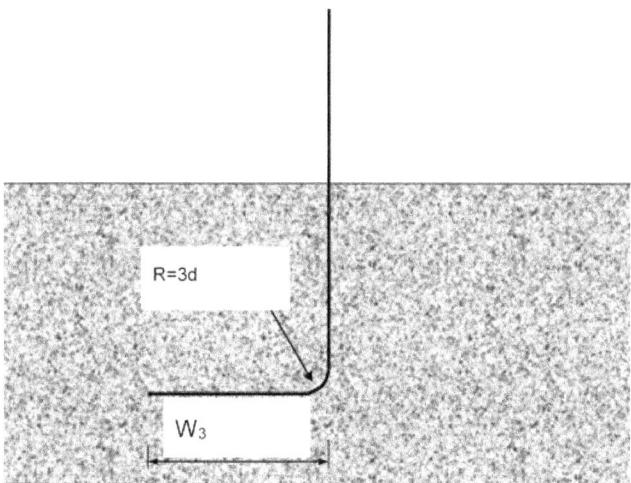

Figure 49 Steel bending detail

Example

Retaining wall as in the following drawing. The height of the retaining wall is 4.0 m and the surcharge load 0.3 t/m²

Solution

The solution is done by excel sheet to obtain this sheet will be through the website. **www.elreedyman.com**

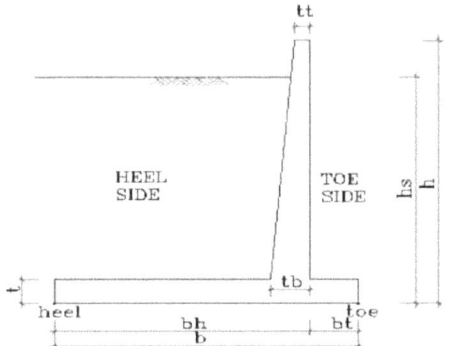

Example 2: INPUT DATA FOR RETAINING WALL DESIGN				
WALL DIMENSIONS				
Height of wall (h) =			4	m
Height of soil (hs) =			4	m
Thickness of the wall top (t_t) =			0.25	m
Thickness of the wall bottom (t_b) =			0.4	m
Width of base (b) =			3	m

Width of toe =			0.75	m
Width of heel (b_h) =			2.25	m
Base thickness (t) =			0.5	m
PROPERTIES OF MATERIAL				
$\gamma c=$			2.5	t/m³
$\gamma s=$			1.8	t/m³
K a=			0.333	
Friction factor, $\mu=$			0.42	
APPLIED LOADS				
Intensity of surcharge load=			0.3	t/m²
CALCULATION OF WEIGHTS				
W1=			2.19	t/m'
W2=			0.66	t/m'
W3=			3.75	t/m'
W4=			0.47	t/m'
W5=			11.66	t/m'
NET WEIGHT=			*18.72*	t/m'
STABILIZING MOMENT=			*32.93*	m.t/m'
EARTH & WATER PRESSURE AT TOP OF FOUNDATION LEVEL				

Force for backfill (F1)=			3.671	t/m'
Surcharge Load(F2)=			0.35	t/m'
EARTH & WATER PRESSURES AT FOUNDATION LEVEL				
FORCE DUE TO BACKFILL(F1)=			4.8	t/m'
FORCE DUE TO SURCHARGE(F6)=			0.4	t/m'
SUM OF SLIDING FORCES=			*5.2*	t/m'
OVERTURNING MOMENT=			*7.2*	m.t/m'
SAFETY FACTORS				
OVERTURNING SAFETY FACTOR=			*4.6*	>2 Ok
SLIDING SAFETY FACTOR=			*1.51*	>1.5 OK
STRESSES ON SOIL				
ECCENTRICITY(e)=	0.125	m		
STRESS AT TOE=	7.8	t/m^2		
EFFECTIVE WIDTH OF BASE=			3	m
STRAINING ACTIONS AT CRITICAL SECTIONS				
STRESS AT SECTION 1=			7.02	t/m2
STRESS AT SECTION 2 =			6.6	t/m2
Moment at Sec.1=			*1.77*	m.t/m'
Moment at Sec.2=			*-3.81*	m.t/m'
Moment at Sec.3=			*4.9*	m.t/m'

Example

The same as the previous example but considering a water table is 3 m above the foundation.

Solution:

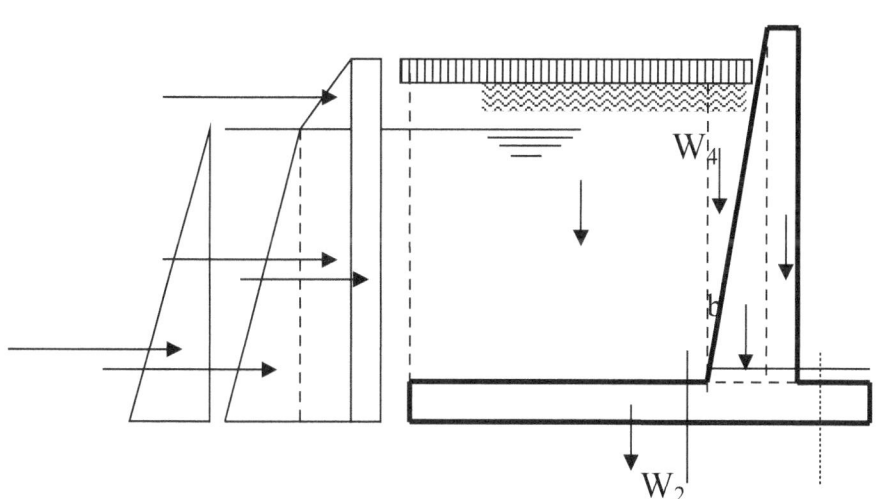

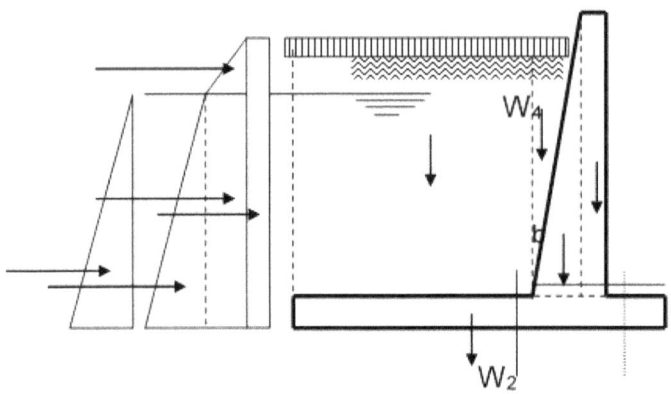

Figure 50 example 2

Example 2: INPUT DATA FOR RETAINING WALL DESIGN				
WALL DIMENSIONS				
Height of wall (h) =			4	m
Height of soil (hs) =			4	m
Thickness of the wall top (tt) =			0.25	m
Thickness of the wall bottom (tb) =			0.4	m
Width of base (b) =			3	m
Width of toe (bt) =			0.75	m
Width of heel (bh) =			2.25	m
Base thickness (t) =			0.5	m
PROPERTIES OF MATERIAL				
$\gamma c =$			2.5	t/m^3

γs=			1.8	t/m³
Submerged soil desnsity, γss=			0.8	t/m³
K a=			0.333	
Friction factor, μ=			0.42	
APPLIED LOADS				
Intensity of surchage load=			0.3	t/m²
WATER HEIGHTS ABOVE FOUNDATION LEVEL				
WATER HEIGHT HEEL SIDE=			3	m
WATER HEIGHT TOE SIDE=			0	m
RESULTS OF STABILITY ANALYSIS , STRESSES AND STRAINING ACTIONS				
CALCULATION OF WEIGHTS				
W1=			2.19	t/m'
W2=			0.68	t/m'
W3=			3.75	t/m'
W4=			0.47	t/m'
W5=			11.66	t/m'
W6=			0	t/m'
UPLIFT=			0	t/m'
NET WEIGHT=			*18.72*	t/m'
STABILIZING MOMENT=			*32.93*	m.t/m'

EARTH & WATER PRESSURE AT TOP OF FOUNDATION LEVEL				
FORCE DUE TO BACKFILL(F1)=			0.3	t/m'
FORCE DUE TO BACKFILL(F2)=			1.5	t/m'
FORCE DUE TO BACKFILL(F3)=			0.83	t/m'
FORCE DUE TO WATER PRESSURE AT HEEL SIDE(F4)=			3.125	t/m'
FORCE DUE TO SURCHARGE(F6)=			0.35	t/m'
EARTH & WATER PRESSURES AT FOUNDATION LEVEL				
FORCE DUE TO BACKFILL(F1)=			0.3	t/m'
FORCE DUE TO BACKFILL(F2)=			1.8	t/m'
FORCE DUE TO BACKFILL(F3)=			2	t/m'
FORCE DUE TO WATER PRESSURE AT HEEL SIDE(F4)=			4.5	t/m'
FORCE DUE TO SURCHARGE(F5)=			0.4	t/m'
SUM OF SLIDING FORCES=			8.2	t/m'
OVERTURNING MOMENT=			10.2	m.t/m'
SAFETY FACTORS				
OVERTURNING SAFETY FACTOR=			*3.23*	>2 Ok
SLIDING SAFETY FACTOR=			*0.96*	< 1.5 Unsafe
STRESSES ON SOIL				
ECCENTRICITY(e)=	0.28546	m		
STRESS AT TOE=		9.8	t/m^2	

STRESS AT HEEL=	2.77	t/m²		
EFFECTIVE WIDTH OF BASE=	3	m		
STRAINING ACTIONS AT CRITICAL SECTIONS				
STRESS AT SECTION 1=	8.0	t/m²		
STRESS AT SECTION 2 =	7.07	t/m²		
Moment at Sec.1=	*2.24*	m.t/m'		
Moment at Sec.2=	*-5.83*	m.t/m'		
Moment at Sec.3=	*6.63*	m.t/m'		

In this case the sliding mode of failure is unsafe so, there are two way of solution to increase the heel to 4.5 m or to construct a stem to contact with the footing to have a passive earth pressure.

6. Soil With Problems

Most the industrial plant specific for oil and gas production plant is to be in remote area so in some cases the location of the plant is in a weak soil so there are many methods of solving these problems depending on the soil type and the soil investigation result.

.6.1 Dynamic compaction

Dynamic compaction (DC), also known as dynamic deep compaction, was advanced in the mid-1960s by Luis Menard. The process depends on dropping a heavy weight on the surface of the ground to compact soils to depths as great as 12.5m.

The purpose of this method is used to reduce foundation settlements, reduce seismic subsidence and liquefaction potential, permit construction on fills, densify garbage dumps, and reduce settlements in collapsible soils.

This method is most effective in permeable, granular soils, but for the cohesive soils tend to absorb the energy and limit the technique's effectiveness.

The ground water table should be at least 1.8 m (6 ft) below the working surface for the process to be effective. For in organic soils, dynamic compaction has been used to construct sand or stone columns by repeatedly filling the crater with sand or stone and driving the column through the organic layer.

This is method is done by using a cycle duty crane to drop the weight, although specially built rigs have been constructed. The crane is typically rigged with sufficient boom to drop the weight from heights of 15.4 to 30.8 m (50 to 100 ft), with a single line to allow the weight to nearly "free fall," maximizing the energy of the weight striking the ground. The weight to be dropped must be below the safe single line capacity of the crane and cable. Typically weights range from 10 to 30 tons and are constructed of steel to withstand the repetitive dynamic forces.

The compaction procedure involves repetitively lifting and dropping a weight on the ground surface. The layout of the primary drop locations is typically on 3.0 to 6.0 m grid with a secondary pass located at the midpoints of the primary pass. Once the crater depth has reached about 1 m, the crater is filled with granular material before additional drops are performed at that location.

It should be considered that this process produces large vibrations in the soil which can have adverse effects on nearby existing structures. It is important to review the nearby adjacent facilities for vibration sensitivity and to document their pre-existing condition, especially structures within 150 m of planned drop locations.

Vibration monitoring during DC is also prudent. Extreme care and careful monitoring should be used if treatment is planned within 200 ft (61.5 m) of an existing structure.

The craters resulting from the procedure are typically filled with a clean, free draining granular soil. A sand backfill can be used when treating sandy soils. A crushed stone backfill is typically used when treating finer-grained soils or landfills.

The depth of influence is related to the square root of the energy from a single drop (weight times the height of the drop) applied to the ground surface. The following correlation was developed by Dr Robert Lucas based on field data:

$$D = k(W \cdot H)^{0.5}$$
(32)

where D is the maximum influence depth in meters beneath the ground surface, W is the weight in metric tons (9 kN) of the object being dropped, and H is the drop height in meters above the ground surface. The constant k varies with soil type and is between 0.3 and 0.7, with lower values for finer-grained soils.

6.2 Vibro Compaction

Vibro compaction was developed in the 1930s in Europe. The process involves the use of a down-hole vibrator (vibroflot), which is to increase bearing capacity, reduce foundation settlements, reduce seismic subsidence and liquefaction potential, and permit construction on loose granular fills.

Applicable soil types: The VC process is most effective in free draining granular soils. The spacing is based on a 165-horsepower (HP) (124 kW) vibrator. Although most effective below the groundwater table, VC is also effective above.

Table 14 Expected improvement and typical probe spacing with vibro compaction

Soil description	Expected improvement	Typical probe spacing (m)
Well graded sand <55 silt, no clay	Excellent	2.7-3.5
Uniform fine to medium sand with <5% silt, no clay	Good	2.3-2.7
Silty sand with 5-15% silt and no clay	Moderate	1.8-2.3
Silty sand >15% or clays and garbage	Not applicable	

The vibrator is lowered into the ground, assisted by its weight, vibration, and typically water jets in its tip. If difficult penetration is encountered, predrilling through the firm soils may also be performed. The compaction starts at the bottom of the treatment depth.

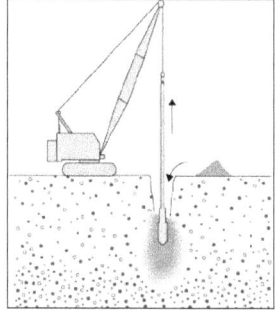 

Figure 51 Vibro system

The vibrator is then either raised at a certain rate or repeatedly raised and lowered configuration, achieving relative densities of 70 to 85%. Treatment as deep as 37 m (120 ft) has been performed.

6.3 Compaction Grouting

Compaction grouting, one of the few US born ground improvement techniques, was developed by Ed Graf and Jim Warner in California in the 1950s. This technique densifies soils by the injection of a low mobility, low slump mortar grout. The grout bulb expands as additional grout is injected, compacting the surrounding soils through compression.

Besides the improvement in the surrounding soils, the soil mass is reinforced by the resulting grout column, further reducing settlement and increasing shear strength. The method is used to reduce foundation settlements, reduce seismic subsidence and liquefaction potential, permit construction on loose granular fills, reduce settlements in collapsible soils, and reduce sinkhole potential or stabilize existing sinkholes in karst regions.

Three main equipment are required to perform compaction grouting, one to batch the grout, one to pump the grout, and one to install the injection pipe. In some applications, ready-mix grout is used eliminating the need for on-site batching.

The injection pipe is typically installed with a drill rig or is driven into the ground. It is important that the injection pipe is in tight contact with the surrounding soils. Otherwise the grout might either flow around the pipe to the ground surface or the grout pressure might jack the pipe out of the ground.

Augering or excessive flushing could result in a loose fit. The pump must be capable of injecting a low slump mortar grout under high pressure. A piston pump capable of achieving a pumping pressure of up to 1000 psi (6.9 MPa) is often required

Compaction grouting is typically started at the bottom of the zone to be treated ground surface and can be terminated at any depth. The technique is very effective in targeting isolated zones at depth. It is generally difficult to achieve significant improvement within about 2.5 m of the ground surface. Some shallow improvement can be accomplished using the slower and more costly top down procedure. In this procedure, grout is first pumped at the top of the treatment zone. After the grout sets up, the pipe is drilled to the underside of the grout and additional grout is injected. This procedure is repeated until the bottom of the treatment zone is grouted. The grout injection rate is generally in the range of (0.087 to 0.175m3/min), depending on the soils being treated. If the injection rate is too fast, excess pore pressures or fracturing of the soil can occur, reducing the effectiveness of the process.

6.4 Wick Drain

The purpose of wick drains is to accelerate the consolidation settlement of soft, saturated clays by reducing the drainage path. A wick drain is a prefabricated drainage strip that consists of a plastic core surrounded by a nonwoven polypropylene geotextile jacket. The geotextile (a filter fabric) allows passage of water into the core that is then pumped out.

A wick drain is installed by enclosing the drain in a tubular steel mandrel and supporting it at the base by an anchor plate. The mandrel is then driven into the soil by a vibratory rig or pushed by a hydraulic rig. At the desired depth, the mandrel is removed, the drain and anchor plate remain in place, and the drain is cut off with a tail about 300 mm long.

Generally, a surcharge program is considered when the site is underlain by soft fine-grained soils which will experience excessive settlement under the load of the planned structure. Using consolidation test data, a surcharge load and duration is selected to preconsolidate the soils sufficiently such that when the surcharge load is removed and the planned structure is constructed, the remaining settlement is acceptable.

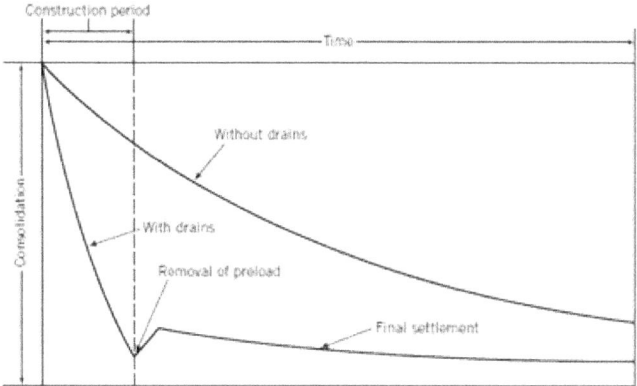

Figure 52 Effect of wick drain with time

The time required for the surcharge settlement to occur depends on the time it takes for the excess pore water pressure to dissipate. This is dictated by the soils permeability and the square of the distance the water has to travel to get to a permeable layer. The PVDs accelerate the drainage by shortening the drainage distance. The spacing are designed to reduce the consolidation time to an acceptable duration. The closer the drains are installed (typically 3 to 6 ft on center) the shorter the surcharge program is in duration.

Procedure: Fill soil is typically delivered to the area to be surcharged with dump trucks. Dozers are then used to push the soil into a mound. The height of the mound depends on the required pressure to achieve the required improvement.

The wick drain typically are in 300 m (1000 ft) rolls and are fed into a steel rectangular tube which is the mandrel from the top. The mandrel is pushed, vibrated, driven or jetted vertically into the ground with a mast mounted on a backhoe or crane. An anchor plate or bar attached to the bottom of the wick drain holds it in place in the soil as the mandrel is extracted. The wick drain will be cut off slightly above the ground surface and another anchor is attached. The mandrel is moved to the next location and the process is repeated. If obstructions are encountered during installation, the wick drain location can be slightly offset.

References:

1. ASTM D1586 - 08a Standard Test Method for Standard Penetration Test (SPT) and Split-Barrel Sampling of Soils

2. ASTM D3441 - 05 Standard Test Method for Mechanical Cone Penetration Tests of Soil

3. ASTM D2573 - 08 Standard Test Method for Field Vane Shear Test in Cohesive Soil

4. ASTM D4428 / D4428M - 07 Standard Test Methods for Cross hole Seismic Testing

5. Seed, H.B. and De Alba, P. (1986). Use of SPT and CPT Tests for Evaluating the Liquefaction Resistance of Sands, Proc., In Situ '86, ASCE, pp.281-302.

6. Peck, R.B., Hanson, W.E., and Thornburn, T.H. Foundation Engineering, second edition, John Wiley and sons, New York, 1996

Index

A

AASHTO 11
aboveground 56
absolute 29
abutments 96
accelerate 123, 124
acceleration 37
accelerometer 37
accept 101
acceptable 124
access 25
accommodate 46
achieve 101, 122, 125
ACI 87
acquire 26
acquisition 48
action 60
advantage 32, 77
aggregates 84
aggressive 9, 84
angle 37, 60, 61, 91
angular 14
API 89
apparatus 25
applied 7, 52, 60, 68, 94, 95, 104, 109, 112, 118
applying 99
approximate 25
approximately 34
arrive 50

arrived 53
assessed 87
assessing 9
assessment 9
assist 56
assisted 120
associated 5, 28
Association 11
assumed 68, 88
assurance 3
assure 37
ASTM 1, 14, 43, 48, 57, 87, 94, 126
at-rest 61
attention 26
Atterberg 11
auger 21, 27, 28
Augering 122
augers 27
Author 2
availability 64
available 22, 26, 57, 69
average 14, 61
avoid 72, 76
axis 31, 48, 53

B

backfill 58, 61, 109, 113, 114, 118
bars 106
basement 23, 96
beam 65, 70, 72-74, 84, 94
bearing 1, 6, 7, 26, 50, 51, 53-55, 59, 62, 64, 66, 67, 70, 74,
    76-78, 87, 88, 90-92, 94, 98, 103-105, 119

bedrock 9, 19, 20
bending 48, 70, 74, 83, 84, 106, 107
body 7, 49
bond 93
borehole 18, 27, 31, 33, 42, 46
boreholes 18-20, 31, 46, 47, 64
boring 17, 21, 22, 24-26, 28-31
borings 6, 17-19, 21-25, 27, 29, 57
borrow 16, 28, 30
boulders 31
breadth 30
break 8
bridge 23, 96
Bryant 6, 7
building 17-19, 25, 30, 74
bulb 121

C

cable 48, 117
cables 26, 31
calcium 85
calculated 43, 54, 100
calculating 7, 89
cantilever 72, 96, 97
cap 92, 93
caps 1, 92, 93
Casagrande 12
cement 82, 84, 85
checklist 4
Chemical 61
chemicals 84
civil 3-6, 9, 10

Class 40
Classes 40
classification 10-12, 14, 28, 31, 63
classifications 10
classified 10
clay 7, 14, 39, 78, 91, 92, 119, 120
clayey 35, 55
clays 18, 19, 120, 123
Coarse 14
coarse-grained 16, 19, 95
coarsely 14
code 54
cohesionless 69, 76
cohesive 1, 11, 26, 41, 53, 55, 69, 89, 90, 117, 126
collapse 6, 33
collapsible 62, 116, 121
column 67, 68, 70-72, 92, 93, 117, 121
compact 116
compacted 21
compressibility 59
compression 47, 59, 121
comprised 7
concept 87
concrete 1-3, 8, 30, 61, 65, 70, 72, 77, 81, 82, 84-87, 91, 93, 98, 99, 101, 105
conductivity 37, 61
confine 96
confinement 33
conical 36
conservative 55
Consistency 11, 13, 35, 41

consolidation 28, 123, 124
constructed 10, 117, 124
construction 4, 6, 9, 10, 25, 31, 56, 59, 62, 64, 81, 83, 102, 116, 119, 121
content 11, 12, 16, 17, 61-64, 84
contract 63, 64
coordinate 57
coordinates 31, 47, 64
core 22, 41, 87, 123
cores 87
CORNER 70
correlation 118
costs 21
CPTU 37, 39
cracks 23
crane 117, 125
cross 1, 46-48, 83, 126
cross-hole 46
cured 83
curve 52, 53, 74
cutting 17, 29
cycle 117
cylinder 29

D

damage 6
dams 25, 30
datum 57
dead 67, 70, 97
debris 9
deep 4, 6, 46, 58, 59, 76, 116, 121
deeper 76

definition 4, 16, 63, 64
deflection 60
deformation 50
degree 3, 16, 68
degrees 40, 41
dense 19, 36, 41, 53, 55
densify 76, 116
density 16, 35, 49, 101
depth 4, 17-20, 24-27, 30-33, 42, 46, 47, 51, 54, 58-60, 62, 70, 74, 78, 94, 117-120, 122, 123
depths 10, 26, 27, 33, 50, 55, 61, 116
designer 72, 74
detail 4, 18, 106, 107
detailed 6, 9, 23, 40
dewatering 62
dikes 30, 63
dimension 65, 98
dimensionless 90, 91
direction 39, 49, 60, 64, 72
directly 27
disadvantages 77
Discuss 62
dissolved 84
distance 47, 74, 124
distances 66, 84
distribute 92
disturb 28, 33
disturbance 28
disturbed 27, 28, 32
documents 63, 64
down-hole 119

drainage 31, 61, 62, 123, 124
drains 31, 123, 124
drill 20, 33, 122
drilled 18, 46, 59, 60, 81, 122
drilling 18, 21, 26, 27, 31-33, 46
drop 117-119
drying 79
dynamic 1, 33, 58, 116, 117

E

earthquake 46, 59, 60, 97
earthquakes 9
earthworks 9
economic 3, 74
economical 6, 9, 66, 72
efficient 9
effort 24
elastic 49
electric 31, 36, 37
electronic 48
electronics 48
element 4, 37, 72, 74, 105
elevations 57, 60
embankment 17, 20
Embankments 63
embedded 106
embedment 18, 59
end-bearing 88
energy 7, 33, 117, 118
engineer 4, 5, 9, 21, 24, 26, 32
environment 7, 9, 28
erosion 23, 58, 63

exploration 10, 21, 23, 26, 27
exploratory 24

F

facilities 46, 60, 62, 63, 118
factor 16, 34, 50, 51, 53, 67, 90, 91, 99, 101, 102, 104, 109, 110, 112, 114
failure 6, 25, 52, 53, 55, 96, 115
failures 6
fall 117
fault 9
fcu 70, 73
fence 8
fill 57, 61-64, 125
filling 73, 81, 102, 117
fine-grained 39, 124
finite 72, 74
flat 72, 74
floor 25
fluid 33, 49
footing 30, 53, 54, 66, 68, 71-74, 115
footings 33, 54, 58, 66, 72
footprint 18
free 85, 117-119
free-fall 33
frequency 25, 33, 50
frequently 29
friction 37-40, 59-61, 77-79, 88-92, 94, 109, 112
function 48, 59
functions 76

I

inclination 40, 41, 47, 63
index 13, 16
infinitive 48
injected 121, 122
injection 121, 122
installation 60, 125
isolated 54, 66, 72, 74, 122

J

jacket 123
jacking 29, 94

K

key 64, 70, 102

L

laboratory 26, 28, 29, 55, 57, 61, 62
landfills 118
lateral 25, 38, 60, 83, 91, 97
layer 30, 77, 78, 91, 92, 117, 124
layers 7, 27, 39, 76
layout 26, 117
length 20, 40, 59, 60, 67, 71, 83, 86, 91, 94
liquefaction 9, 62, 116, 119, 121, 126
loaded 6
loads 7, 18, 25, 59, 60, 76-78, 92-94, 109, 112
locations 18, 26, 57, 117, 118
Long-term 59

M

management 3
manufacturing 17, 30

map 58
Marine 3
materials 3, 57, 64, 70, 77
measurement 38, 39, 42, 48
measurements 37-40, 59
mechanics 7, 88
mechanism 51
multistory 17, 30
multi-story 74

N

non-cohesive 52, 53
nonstructural 64
non-uniform 49, 50

O

obstruction 31
onsite 39, 52
organic 7-9, 117
Organization 24
overburden 28, 34, 39, 90
overlap 71, 74
overturn 99
overturning 1, 98-101, 110, 114

P

Particle 14
particles 14, 49
passive 61, 102, 115
pavement 62
percentage 15
phenomena 46
piezo 37

piezo-transducer 37
pipe 27, 29, 81, 82, 86, 87, 121, 122
pipes 31, 46, 81
planned 25, 118, 124
planning 21-23, 29
plastic 11, 12, 17, 123
Plasticity 13
plate 6, 50-52, 54, 55, 123, 125
plates 48, 54
Poisson's 58
Portland 84
pouring 82
pre-bored 32
preliminary 1, 4, 23, 97, 98
pressures 10, 61, 70, 109, 113, 123
principals 25
probe 37, 48, 119
probes 46
project 4, 5, 9, 21, 23, 29, 46, 56, 96
projects 2, 3, 21, 33, 36-38
pump 121, 122
pumped 122, 123
push 29, 51, 125
PVC 46
PVDs 124
P-wave 49
P-Y 60

Q

quantifies 16
quantity 63
question 90

quick 95
quite 53

R

range 46, 87, 117, 122
ratio 16, 38, 39, 49, 58, 62, 84, 91
ready-mix 121
receiver 46, 47
reclamation 33
recommendation 4, 59, 65
reflect 54
reinforcement 70, 72, 74, 87, 92, 106
reinforcements 83
repair 3
reports 4
rig 34, 83, 122, 123
rigged 117
ring 37
ringwall 58
risk 9, 58
risks 9
rock 9, 14, 19, 63, 76, 77
rocks 8
rotation 43
Rule-of-thumb 41

S

sample 12, 24, 25, 27-29, 51
samples 23, 24, 26-29, 32
sampling 1, 25, 27, 29, 33, 57, 126
saturated 16, 123
saturation 16, 37
seamless 29

seamless-steel 81
sensitive 76
sensors 37, 94
settlement 8, 21, 23, 30, 33, 50, 52-55, 59, 60, 63, 76, 94, 95, 121, 123, 124
sheet 5, 69, 108
shell 81, 82, 86, 87
siliceous 84
silt 14, 39, 119, 120
silts 33
silty 17, 35, 64, 120
skeleton 49
slab 62, 72, 74
sleeve 37-40
sliding 1, 98, 101-103, 110, 114, 115
slope 20, 59, 62, 63
slopes 63
slump 121, 122
slurry 60
soft 7, 18, 19, 35, 36, 41, 123, 124
soils 1, 6, 7, 16, 18, 19, 25, 26, 49, 52, 53, 55, 61, 62, 76, 92, 95, 116-122, 124, 126
solid 17, 50
solution 74, 91, 107, 108, 111, 115
solutions 60, 61, 64
Splices 81
stability 1, 25, 63, 74, 98, 99, 101-103, 113
stabilization 33, 63
stabilize 121
stabilizing 31, 109, 113
stem 106, 115

stiff 19, 35, 36, 42, 91, 92
stiffness 33, 92
storage 46
stored 29
story 17
strain 94
strap 72-74
strata 26, 57, 76
strategy 2
stratification 10, 23, 37, 39, 40
stratifications 28
Stratigraphy 56
stratum 19, 55, 78
stress 18, 25, 33, 34, 39, 48, 71, 83, 87, 91, 98, 110, 114
stressed 30
stresses 68, 83, 84, 93, 94, 104, 105, 110, 113, 114
striking 117
strip 18, 123
stripping 64
structural 2, 5, 61, 64, 76
structure 2, 4-6, 9, 10, 17-19, 21, 23, 25, 29, 30, 81, 97, 118, 124
structures 2, 3, 6, 10, 23, 25, 26, 54, 55, 76, 118
sub-base 62
subgrade 58, 62-64
subgrades 63
subsoil 9, 21, 25, 26
sub-soils 59
sulpher 84
support 7, 26, 66, 72
supporting 123

supports 10
surchage 112
surcharge 107, 109, 110, 113, 114, 124, 125
surcharged 125
Surfaces 49, 50
Surfacing 62
sustain 76
Swell 16
Swelling 62

T

technical 3-5
techniques 121
Telecom 3
tensile 83
tension 93
Terzaghi 7
test 1, 12, 21, 23, 26, 29, 32, 33, 37, 41-48, 50, 52, 54, 55, 57, 61, 93-95, 124, 126
tested 33
testing 1, 3, 26, 29, 32, 42, 48, 59, 62, 126
tests 1, 4, 11, 12, 21, 28, 36, 42, 55, 57, 58, 94, 126
theoretical 7, 33
theory 7, 87, 89
timber 77, 79, 80, 87
time 5, 7, 8, 43, 48, 59, 124
tip 17, 36, 37, 42, 60, 77, 78, 82, 120
topography 23, 56
topsoil 57, 64
trunks 79

U

ultimate 50, 53, 55, 59-61, 78

underground 26, 57, 84
underside 122
undrained 42, 51, 90
uplift 60, 113

V

variations 18, 57
vibro 1, 119, 120
volume 16, 17

W

wall 1, 4, 8, 20, 26, 29, 91, 96-99, 101-103, 106-108, 112
walls 25, 96, 97, 99
washed 82
website 108
weight 15, 16, 41, 51, 61, 84, 97, 100, 109, 113, 116-120
welded 81
welding 81
wood 79, 80

X

x-y-z 48

Y

Yearly 61
years 8
yet 7, 37, 38
yield 21
York 126

Z

zone 62, 122
zones 30, 122

www.ingramcontent.com/pod-product-compliance
Lightning Source LLC
Chambersburg PA
CBHW050004230526
45465CB00003BB/1248